AF413263

BETWEEN NATURE AND SOCIETY

Biographies of Materials

A World Scientific Encyclopedia of the Development and History of Materials Science

Published

Between Nature and Society: Biographies of Materials
    edited by Bernadette Bensaude-Vincent

Between Making and Knowing: Tools in the History of Materials Research
    edited by Joseph D Martin and Cyrus C M Mody

A World Scientific Encyclopedia of the Development and History of Materials Science – Vol. 2
Editor-in-Chief: Arne Hessenbruch

BETWEEN NATURE AND SOCIETY

Biographies of Materials

Editor

Bernadette Bensaude-Vincent

University of Paris 1 Pantheon-Sorbonne, France

NEW JERSEY · LONDON · SINGAPORE · BEIJING · SHANGHAI · HONG KONG · TAIPEI · CHENNAI · TOKYO

Published by

World Scientific Publishing Co. Pte. Ltd.
5 Toh Tuck Link, Singapore 596224
USA office: 27 Warren Street, Suite 401-402, Hackensack, NJ 07601
UK office: 57 Shelton Street, Covent Garden, London WC2H 9HE

Library of Congress Control Number: 2022934990

British Library Cataloguing-in-Publication Data
A catalogue record for this book is available from the British Library.

A World Scientific Encyclopedia of the Development and History of Materials Science
BETWEEN NATURE AND SOCIETY
Biographies of Materials

Copyright © 2022 by World Scientific Publishing Co. Pte. Ltd.

All rights reserved. This book, or parts thereof, may not be reproduced in any form or by any means, electronic or mechanical, including photocopying, recording or any information storage and retrieval system now known or to be invented, without written permission from the publisher.

For photocopying of material in this volume, please pay a copying fee through the Copyright Clearance Center, Inc., 222 Rosewood Drive, Danvers, MA 01923, USA. In this case permission to photocopy is not required from the publisher.

ISBN 978-981-125-174-0 (hardcover)
ISBN 978-981-125-175-7 (ebook for institutions)
ISBN 978-981-125-176-4 (ebook for individuals)

For any available supplementary material, please visit
https://www.worldscientific.com/worldscibooks/10.1142/12709#t=suppl

Desk Editor: Joseph Ang

Typeset by Stallion Press
Email: enquiries@stallionpress.com

About the Contributors

Hervé Arribart has shared his career between academia and industry in France. He began as a Centre National de la Recherche Scientifique (CNRS) researcher in solid-state physics. He soon moved to Elf Aquitaine Company where he worked on sensors and instrumentation for medical diagnosis, in tight collaboration with DuPont. Then, he joined the Saint-Gobain Company as a research team leader in the field of materials surfaces and interfaces. In 1990, he founded the joint CNRS/Saint-Gobain laboratory, a public–private scientific joint-venture. In parallel, he was teaching quantum mechanics and solid-state physics at Ecole Polytechnique. As the scientific director of Saint-Gobain in 2003, he was in charge of the coordination of research programs, of the management of exploratory research and academic collaborations. In 2009, he moved back to an academic position, as a professor in physics and energetic systems at the Ecole Supérieure de Physique et de Chimie Industrielle.

Bernadette Bensaude-Vincent, Professor (em) at Paris1 Panthéon-Sorbonne University (France), is a historian and philosopher of science and technology. Her research explores technosciences in general, and materials science in particular. Among her recent publications, she authored *Les Vertiges de la technoscience* (2009), *Carbone: ses vies, ses oeuvres.* (S. Loeve co-author 2018) and *Temps-paysage. Pour une écologie des crises* (2021). She co-edited with Loeve S., Normann A., Schwarz A. *Research Objects in their Technological Setting* (2017), and with S. Loeve,

X. Guchet *French Philosophy of Technology. Classical Readings and Contemporary Approaches* (2018). She is a member of the French Academy of Technology and recipient of the 2021 Sarton Medal.

Harry Bernas is Research Director (em.) at CNRS and University of Paris-Saclay, Orsay, France. He has authored some 250 papers on condensed matter physics and materials science topics, and held management positions in the French and European research systems. A long-term interest in societal interactions of science has led him, over a decade, to study the impact of politics on the aims and practice of science and technology. Nuclear energy, being emblematic in this regard, is the main focus of his current work. He can be reached at harry.bernas@u-psud.fr.

Soraya Boudia is an STS scholar, Professor at University of Paris. Her work explores the role of science in politics and policy, particularly the transnational government of health and environmental risks. She has edited with N. Jas *Powerless Science? Science and Politics in a Toxic World* (Berghann, 2014) and *Toxicants, Health and Regulations Since 1945* (Pickering and Chatto, 2013). She is co-editor with A. Creager, S. Frickel, E. Henry, C. Reinhardt, J. Roberts, of *Residues: Thinking Through Chemical Environments* (Rutgers University Press, December 2021).

Kate Brown is Professor of Science, Technology and Society at the Massachusetts Institute of Technology. She is the author of several prize-winning histories, including *Plutopia: Nuclear Families, Atomic Cities, and the Great Soviet and American Plutonium Disasters* (Oxford, 2013). Her latest book, *Manual for Survival: A Chernobyl Guide to the Future* (Norton, 2019), translated into nine languages, is a finalist for the 2020 National Book Critics Circle Award, the Pushkin House Award and the Ryszard Kapuściński Award for Literary Reportage.

Hasok Chang is the Hans Rausing Professor of History and Philosophy of Science at the University of Cambridge. He received his degrees from Caltech and Stanford, and has taught at University College London. He is the author of *Is Water H_2O? Evidence, Realism and Pluralism* (2012) and *Inventing Temperature: Measurement and Scientific Progress* (2004). He is also co-editor (with Catherine Jackson) of *An Element of Controversy: The Life of Chlorine in Science, Medicine, Technology and War* (2007), a collection of original work by undergraduate students at University College London. He is a co-founder of the Society for Philosophy of Science in Practice (SPSP) and the Committee for Integrated History and Philosophy of Science.

Philippe Dillmann is a Research Director at the CNRS, an engineer in materials science and a specialist in the study of ancient metals. For more than 20 years he has been conducting research in archaeological sciences aimed at understanding the techniques and know-how for the production of metals, their modes of exchange, their use and reception in ancient societies. For the Middle Ages, his work focuses on the role of metal in the construction of the great Gothic monuments. His research also focuses on the alteration and conservation of metal heritage objects, including the prediction of the long-term behavior of materials used by contemporary societies. He received the 2020 silver medal of the CNRS.

Florence Hachez-Leroy, Associate Professor of Modern History at Artois University and Researcher at the *Centre de recherches et d'études Histoire et société*, works on the history of modern industrial materials with a focus on the economic, technical, social, cultural and heritage aspects of their entrepreneurial history. Her research addresses the role of international cartels in the aluminum sector and their impact on the market. She is currently PI of a national research program called "Aluminum, Architecture and Heritage, 20–21st century" (Archipal), 2019–2023. Her most recent book, *Menaces sur l'alimentation* (2019), focuses on environmental history, particularly the regulation of food additives and their relation to public health, in Europe and the United States.

Stephen L. Harp is a distinguished Professor of History at the University of Akron. He is the author of *Learning to Be Loyal: Primary Schooling as Nation Building in Alsace and Lorraine, 1850–1940* (Northern Illinois University Press, 1998); *Marketing Michelin: Advertising and Cultural Identity in Twentieth-Century France* (Johns Hopkins University Press, 2001); *Au Naturel: Naturism, Nudism, and*

Tourism in Twentieth-Century France (Louisiana State University Press, 2014); *A World History of Rubber: Empire, Industry, and the Everyday* (Wiley-Blackwell, 2016); and *The Riviera, Exposed: An Ecohistory of Postwar Tourism and North African Labor* (Cornell University Press, forthcoming in 2022).

Emmanuel Henry is a Professor of Sociology at Université Paris-Dauphine, PSL University in France and a former member of the School of Social Science at the Institute for Advanced Study, Princeton, New Jersey. He is currently working on the links between scientific knowledge, ignorance, expertise and public policy, in the fields of environmental and occupational health. He is the author of *Amiante: Un scandale improbable* [Asbestos: an Unlikely Scandal] (Presses universitaires de Rennes, 2007); *Ignorance scientifique et inaction publique* [Scientific Ignorance and Public Inaction] (Presses de Sciences Po, 2017) and more recently *La fabrique des non-problèmes* [The Construction of Non-Problems] (Presses de Sciences Po, 2021).

Eric Le Bourhis is Professor in the Physics Department at Poitiers University (France). He gained his PhD at Paris VII University in 1994, then joined Saint Gobain R&D as an engineer where he developed expertise in glass surfaces and coatings. He joined Poitiers University in 1998, where he has pursued an activity to promote coatings in close collaboration with glass industrial manufacturers. He covered both aspects of glass science and technology in a book published recently (Glass Mechanics and Technology, Wiley VCH, 2014).

Sacha Loeve is Associate Professor of Philosophy of Science and Technology at the University Lyon 3 Jean Moulin and member of the Lyon Institute of Research in Philosophy (IRPhiL). He received his degrees from Paris Nanterre University. He is interested in the practices, narratives and knowledge of various technoscientific objects from molecular machines to nanocarbons and biomaterials, which he studies in the field to draw a different portrait of our contemporary material culture. He has co-edited *Research Objects in their Technological Setting* (Routledge, 2017), *French Philosophy of Technology* (Springer, 2018) and co-authored a biography of carbon (*Carbone. Ses vies, ses oeuvres*, Le Seuil, 2018) with Bernadette Bensaude-Vincent.

M. Grant Norton is Dean of the Honors College and Professor of Mechanical and Materials Engineering at Washington State University. Norton obtained his PhD in Materials from Imperial College and was a postdoctoral fellow at Cornell University. He has held visiting appointments at Wright-Patterson Air Force Base, Oxford University, the Chien-Shiung Wu Honors College at Southeast University and as a Global Faculty Fellow at Tecnológico de Monterrey.

Judith Rainhorn is a Professor in Modern History at University Paris 1 Panthéon-Sorbonne and a Senior Researcher at the Maison Française in Oxford. Her research interests include the history of medicine and health, environmental and labor history in France, Europe and the United States, 19th–20th c., on which she has published extensively. Her most recent book, *Blanc de plomb. Histoire d'un poison legal* (Presses de Sciences Po, 2019) was awarded the 2020 François-Bourdon Academy Prize, the 2020 Prescrire Prize and the special mention of the OCIRP-Francis Blanchard-ILO 2020 prize. In 2020, Judith received the *Fondation des Sciences Sociales* Award for excellence in research.

Pierre Teissier is Associate Professor in Epistemology, History of Science and Technology at the University of Nantes (France). His research interests are linked to the history of materials sciences, especially solid-state chemistry, and to the history of energy in the modern period. His current research deals with the problem of energy storage in Western Europe and Northern America from the 1870s onward.

Frédéric Thibault-Starzyk is Directeur de Recherche in the French National Centre for Scientific Research (CNRS). He is a chemist, working in (*operando*) spectroscopy in catalysis. He designed experimental methods for understanding surface chemistry for a new energy efficient and sustainable chemistry. After a fellowship in Churchill College Cambridge, Fred Thibault-Starzyk became the director of the Catalysis and Spectrochemistry laboratory in Caen (2011–2016) and of the Maison

Française (2016–2020) in Oxford, where he was elected Senior Associate in Pembroke College. He was the President of the French Zeolite Association between 2009 and 2014.

Henri Van Damme is a physical chemist and material scientist. Now retired from the French higher education system and working as an independent consultant, he has spent the last years (2014–2017) as visiting professor in the Civil & Environmental Engineering Department (CEE) of the Massachusetts Institute of Technology (MIT) and as member of the MIT–CNRS (French National Center for Scientific Research) joint research unit on multiscale materials for energy and the environment, on leave as emeritus professor from the Ecole de Physique et Chimie Industrielles de Paris (ESPCI-Paris) where he has been professor of thermodynamics and materials science since 1999. Henri Van Damme has devoted most of his career to the physical chemistry and physics of geo-materials like glass, clays, cement, concrete or source rocks, with applications in the field of construction, catalysis, energy, smart materials or biology. He is also interested in architecture, conservation, urban sciences and counter-intuitive teaching methods.

Catherine Verna is Professor of Medieval History at the University of Paris 8 Vincennes-Saint-Denis and a member of the CNRS. She co-leads the CNRS Research Group: *Techniques et production dans l'histoire*. Her research concerns the history of medieval industry, both in its economic and social dimensions and in its technical aspects. She is particularly interested in the metallurgical industry from the 12th to the 16th centuries in the western Mediterranean. She has directed the Casa de Velazquez research program on rural enterprise (ERMO). She has published *Le temps des moulines. Fer, technique et société dans les Pyrénées centrales (13th–16th)*, Publications de la Sorbonne, 2001 and *L'industrie au village. Essai de micro histoire*, Les Belles Lettres, 2017.

Contents

Introduction

Bernadette Bensaude-Vincent

This volume, part of the WSPC Encyclopedia of the Development and History of Materials Science, aims at demonstrating the importance of materials in our life and history. Despite their ubiquity and remarkable diversity, materials are quasi-invisible in our daily life because most of us handle things — artifacts or natural products — rather than materials *per se*. Yet, they have been our first experience of the world as infants and then as major constituents of the world we live in — they deserve our attention.

To show how materials matter, this collective volume develops a multidisciplinary perspective. Some chapters are written by materials scientists, others by historians, philosophers, or science and technology scholars. As a philosopher, I myself became more interested in the stuff things are made of than the transcendental ego, or the mind/body problem. I developed an immoderate fondness for old and new materials, and curiosity for the apparently endless process of substitution of materials and their impact on society, economics and culture.

To introduce the reader into the world of materials, I first characterize the hybrid identity of materials. In a second section, I examine how materials, traditionally in the hands of artisans, builders or engineers, became objects of science. The third section explains the choice of a biographical approach to materials. And the final one presents the organization of the book and the ways to use it.

1. Partners of Technological Adventures

Cyril Stanley Smith, a metallurgist and author of several books on the history of materials, pointed out that archeological ages are named by the material used: stone age, bronze age and iron age [1]. His message was that materials are much more than the microstructures that physicists have been exploring for decades with instruments such as X-ray diffraction, Transmission Electron Microscopes and more recently with Atomic Force Microscopes. Materials are not just atomic structures determining a set of properties and behaviors. They have a social life, and a market life that captures the attention of nations worldwide, since materials have major impacts on our welfare and can affect international peace and security.

Materials such as metals participated in the long process of humans' domestication of the world and adaptation to their environment. In modern times, they contributed to the domination and emancipation of humans from nature. Indeed, historians relying on written sources are less inclined than archaeologists who have to work with artifacts only, to point out the material aspects of the societies under scrutiny. However, they remark that steel massively used in construction and for new technologies such as railways has shaped the industrial world in the 19th century, or that plastics deeply transformed our life styles in the 20th century [2]. Historians cannot ignore that many colonial empires were built to seize and control the mineral resources used for the industrial needs of European nations. The uneven geographical distribution of raw material resources still fuels 21st century geopolitics. Due to the fast population growth and economic development of emerging countries, the Western-dominated order of the world is shifting toward a multipolar world, where international trade flows play a key role [3]. Fears of disruption in the supply of raw materials are a major concern at the highest political levels especially in fast developing countries.

So avid is our contemporary exploitation and consumption of materials that our technology development impacts on the history of the Earth: it leads to the Anthropocene, the new geological era when humans act as geological forces. Significantly, to celebrate the 150th anniversary of the discovery of the periodic system by Dmitri Mendeleev in 2019, the European Chemical Society (EuChemS) issued an unfamiliar table of

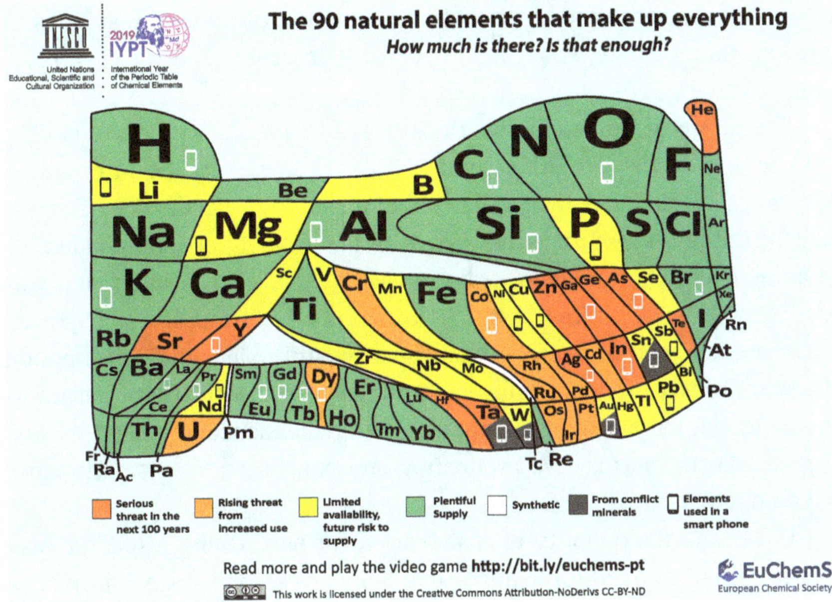

Figure 1. Periodic system of the 90 elements used in technologies. Colors indicate their level of abundance or depletion.

Source: EuChems, https://www.euchems.eu/euchems-periodic-table/.

elements (Figure 1). This landmark table distorts the conventional rectangular graphic presentation that you find in all chemistry laboratories and textbooks around the world, in order to point out the scarcity of elements.

Chemical elements, the building blocks of all materials, are not defined by their atomic numbers, weights, and properties only, they are primarily viewed as resources for human technologies (hence the elision of the lanthanides and actinides). In other terms, they feature as hybrid entities belonging to nature and culture. Unlike the rigid chart of the periodic system, this updated version is flexible and is evolving over time according to the fluctuations of humans' uses of material resources. David Cole-Hamilton, EuChemS vice-president and Emeritus Professor in Chemistry at the University of St Andrews, insisted on the political message of this provocative table. It is meant to draw people's attention to the lifespan of elements and to the risks of depletion of mineral resources. It is meant to encourage sober manufacturing processes and improve recycling practices.

The future of human civilizations heavily depends on the availability of materials. In the late 20th century, it could seem that the Information Age would result in a dematerialization of technology. On the one hand, the miniaturization in electronic industry, with the capacity of semi-conductors doubling every 18 months according to Moore's law, suggested that it was possible to ever "get more from less". On the other hand, it was expected that computers and digital communication would reduce the consumption of material inscriptions of information. However, it turned out that information technologies did in fact increase our dependence on materials, due in particular to the massive production and consumption of smartphones and computers.

To be sure the quantity of matter used for performing a specific task is decreasing. It is said for instance that while 7000 tons of steel have been used to build up the Eiffel Tower in the 1880s, 2000 tons would suffice nowadays, given the higher quality of the steels manufactured. This reduction results from a better control of the inner structure of materials.

The emergence of Materials Science and Engineering has enhanced the imaginary of mastery of materials by increasing materials transparency. When you know everything you can about the materials you use, you can make the best selection for your projects, and for the environment. And if you don't find the optimal material among the list of available ones, you can even mobilize all the resources of cutting-edge research and development to design a new material that fits your demands. In the mid-20th century, material scientists were asked to design materials for the specific performances required for space and aircraft industry. Materials with never-seen-before properties such as the lightness of plastics, the strength of steel and the resistance of ceramics have been specifically designed for rockets or missiles. These artificial products are close to art creations rather than simple commodities. As materials by design no longer are prerequisites imposing constraints on manufacturing processes, they generate a sense of freedom and of overtaking nature. The coming era of materials by design promises to release our technological dependence on nature, to reduce the constraints and limitations to technological innovations imposed by the materiality of artifacts. Moreover, since materials by design use less quantity of matter and increase the effectiveness of materials, material scientists have also promised to solve environmental issues. Ken Geiser, for instance, argued that the design of materials for specific industrial uses is the best way to avoid environmental damage [4].

The champions of nanotechnology went even further in the economy of promises. The design of materials from bottom-up would make things with no fall-off waste and less use of toxic chemicals. "Shaping the world atom by atom", the slogan of the US National NanoInitiative in 2000, was made to capture the popular imagination. The hype surrounding the emergence of nanotechnology came from scientific institutions themselves. In particular Mihail C. Roco, Senior Advisor for Science and Engineering at the National Science Foundation (NSF), self-consciously exaggerated the anticipated benefits of nanotechnology and the potential disruption that would result from its convergence with biotechnology, information technology and cognitive science. Materials were bound to become self-sustaining and self-repairing systems that would abolish all constraints [5].

However, the advances in materials science and converging technologies did neither alleviate our dependency on materials nor fix environmental issues. Although they may have significantly *decreased the quantity of matter* used in the molecular manufacture of a number of products, materials by design did in fact increase *the qualitative diversity* of materials required for manufacturing purposes. The demand for more and more elements is especially high in information technology (screens, smartphones) and energy technology (solar panels and rotor blades of wind turbines) as shown in the diagram (Figure 2). Our technological system is more and more entangled with an increasing number of materials. The dematerialization of technology is an illusion, a dream encouraged in Western countries by the global economy system that prevailed over the past decades. As a significant part of the world industrial production has been massively transferred to Asia where manpower is cheaper, most chemicals are sold, used and released in Asia (169% of the chemicals used in the United States and Europe combined) [6]. From 1900 to 2015, the global amount of fossil and mineral materials extracted has increased by a factor of 12. Material flow studies indicate that the global material demand rose in the 1980s and accelerated in the early 21st century [7].

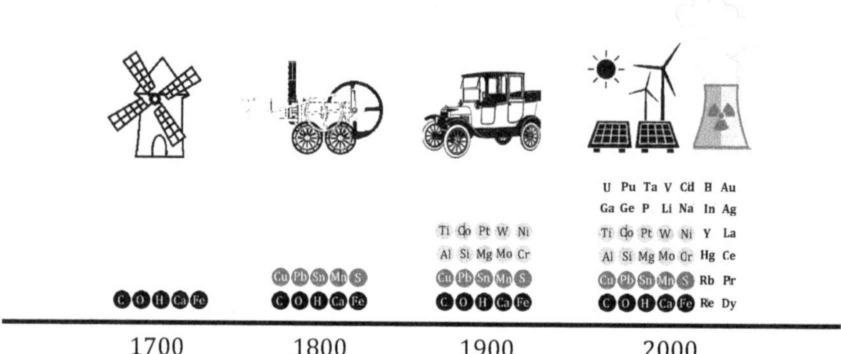

Figure 2.   Elements widely used in energy technology.

Source: Adapted from https://www.weforum.org.

The changing patterns of consumption of raw materials generate new tensions in geopolitics. Indeed, the scarcity of resources is not new in history, over the past decades however it has significantly impacted the global economic system of trade and investment. While Europe imports all of its material and energy resources, China has developed a pragmatic business model for securing the materials supply needed for its economy either by maximizing the benefits of its own resources or by gaining access to materials reserves in various continents [3]. The quest for materials thus remains the main driver of the geopolitical order.

2. Materials in the Plural

Given the importance of materials in human affairs, it is worthwhile to make a point of conceptual clarification to specify what we are talking about. The two terms "matter" and "material" stem from the same Latin substantive *materies,* based on the root *mater* (mother) that conveys the idea of a creative and feminine power. Originally, *materies* referred to timber, more precisely to the core of the tree-trunk (matrix): the part presenting long regular fibers being the most suitable for carpentry [8, p. 16]. Timber is the archetype of both matter and material. However, the substantive 'material' is a hybrid notion referring to an alliance between natural beings and technological needs, "matter" is an abstract and generic notion. Materials are varieties of matter viewed as useful for human purposes.

Unlike matter, materials have unique properties. Timber is carefully selected, chosen by the craftsman for making a violin or a piece of furniture. In the old days when shipbuilding for the merchant marine was a central activity, Dutch shipwrights used to make long journeys to visit pinewoods in specific areas in order to select the most suitable tree for the mast of the ship to build. And more recently, following the fire of the medieval cathedral Notre-Dame-de-Paris, hundreds of old oak trees have been chased and cut for repairing the roof structure. Trees are extracted from nature, separated from their natural environment and turned to timber in relation to a construction project.

The abstract notion of matter came out of a gradual semantic shift from the reference to a definite part of a tree, object of choice for specific technological purpose to timber in general; then from wood to all materials employed in human arts; and finally from craftsmen's materials to the substance of all bodies in general. The modern concept of matter thus appears as the result of a process leading from a singular entity to an abstract notion that 17th century physicists defined by properties such as extension and inertia.

The emergence of modern science deprived the Aristotelian notion of "substance" that is common to all natural and artificial bodies from individual features. René Descartes, for instance, treated color, odor and smell as secondary or accidental qualities and reduced matter to its geometrical extension. Robert Boyle built natural philosophy (the science of nature) upon the notion of a "catholic" or universal matter. Moving beyond the multiplicity and the variety of individual and phenomenological substances enabled natural philosophers to formulate laws of nature. Natural laws — even approximate ones — are universally valid statements. They apply to the behavior and properties of all entities, in all situations, without exception. In this context, the notion of materials being linked to individual features and to human interests was an epistemological obstacle that modern science had to overcome [9].

Chemists, by contrast, have always been concerned with individual substances although they admitted the law of conservation of the quantity of matter in chemical reactions. "No natural body consists of matter *per se*", this remark made by the French chemist Gabriel François Venel in 1765 points to the difference between the generic notion of matter and the individual features of material bodies that are of interest for chemical technologies and engineering [10, pp. 375a–376b].

Galileo wrestled with materials in his dialog on *Two New Sciences*. Thanks to his frequent visits to the Arsenal in Venice, Galileo was familiar with engineering problems. He noticed that the "master" workers knew by experience that they could not apply the same rules and use the same devices for building small ships and big ships. The natural philosopher eager to apply geometry to all solids was tempted to consider this opinion as a popular prejudice since the properties of geometrical figures — the circle, the cone or the pyramid — are always independent from their sizes. From their geometrical point of view, solids should have retained their properties whatever their spatial dimensions. Similar shapes should present similar behavior. However, that is obviously not the case with materials. Standard geometrical reasoning does not apply to a wooden piece because the strength of all material structures is size-sensitive. There is no proportionality between the quantity of matter and the strength of the material structure. Enlarging the size of a material structure affects its mechanical properties. Galileo clearly understood that not only are materials unique, but for each material structure there is an optimal dimension that allows it to maintain itself: "[N]ature cannot produce a horse as large as twenty ordinary horses or a giant ten times taller than an ordinary man" [11, p. 4]. Galileo emphasized that nature's material structures are characterized by a kind of mutual adaptation between their size and the stuff they are made of. He nevertheless managed to initiate a science of materials, which is not based on geometrical reasoning but on rules of proportionality. The proportionality constant enabled engineers to determine and predict the strength of a wooden structure according to its size. However, Galileo cautiously confined himself to comparing beams of different sizes for the same material. He did not venture into comparisons between a steel pole and one made of wood. His "new science" coped with size-sensitive properties but could not seize the structure-sensitive properties that enable comparisons between materials.

According to the old scholastic saying — "there is no science but of the general". How was it possible to invent a science of materials in the plural, a science dealing with such diverse things as pure silicon and cement, paper and semi-conductors, iron and nanocarbons? How to make a genre out of such heterogeneous things? A science of materials presupposes a generic notion of materials allowing comparison between materials.

Reasoning on ideal materials was a first significant step toward materials in the plural. Mathematicians such as Leonhard Euler and Daniel Bernoulli in the 18th century, and Augustin Cauchy, Denis Poisson, Charles Navier in the 19th century, treated material structures as homogeneous and pure deformable continua, endowed with mass and mechanical properties such as shear modulus, elasticity and compressibility. They were able to distinguish various classes of materials (elastic solids, viscous fluids) and to predict their responses to applied forces. Yet, their equations worked only for ideal materials, and dealt with their mechanical properties exclusively. This mathematical science was built on iron and steel. Unlike the wood structures studied by Galileo, iron is a homogeneous and isotropic material that allows mathematical treatment with differential equations. Iron thus became a model material and a driving force for a scientific approach to materials [12]. Metallurgy acted as a nucleus for the growing of materials science when it was transformed by the investigation of the microstructure of metals with X-ray diffraction techniques in the early 20th century. The determination of microstructure gave access to notions such as crystal lattices, dislocation, defect, that provided a key for connecting crystalline microstructures with the macroscopic properties of metals. Models and theories of structure-sensitive properties elaborated by solid-state physicists could then be used for designing new materials [13].

However, the relation between structure and properties is only one part of the construction of the generic notion of materials. It requires that structure and properties be coupled with functions or performance. It is with respect to functionalities that one compares such dissimilar stuff as wood, concrete, paper, polymers, metals, semi-conductors and ceramics. They are all materials *for*…. And they can be substituted for one another: for instance timber can be replaced by concrete or metal for building construction; wood has been replaced by metal and then carbon fiber composites for making skis, tennis rackets, bicycles and so on. The competition between candidates for one application contributed to the emergence of materials science. The plural entity 'materials' first appeared in the language of science policymakers in the USA, under the auspices of a bottleneck for advances in space and military technologies [14]. The Cold War period favored investments in interdisciplinary initiatives to design high-performance composite materials for missiles and rockets, which were later adapted to civil applications such as aeronautics or sports. With the expansion of composites the plural was integrated in one single material.

Still a generic notion of materials did not emerge until the 1980s when market competition and environmental concerns put new demands on material scientists. Materials research was re-oriented toward engineering to meet industrial needs. Following the wave of emphasis on fundamental aspects in the 1950s and 1960s, which wiped away the empirical approaches of the past, a new wave came which praised engineering skills. The materials generic perspective was driven by a general concern for minimizing risk and cost and by a need for controlling all aspects of the business. The linear model — from basic science to industrial applications — was subverted and replaced by a system approach. For the purpose of materials teaching, this new thought style has been imagined with the help of a tetrahedron indicating that four variables have to be simultaneously taken into consideration in the design or selection of materials: structure, properties, performances and process (Figure 3). The generic concept of materials, an abstract notion, has thus been defined by a set of relations between these four parameters. However, the materials

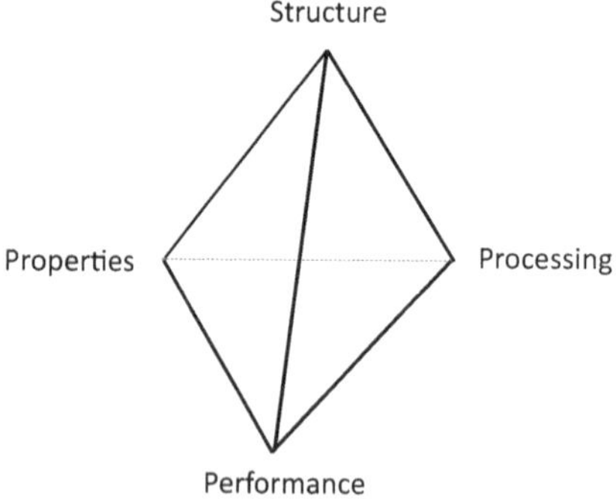

Figure 3.   Materials tetrahedron.

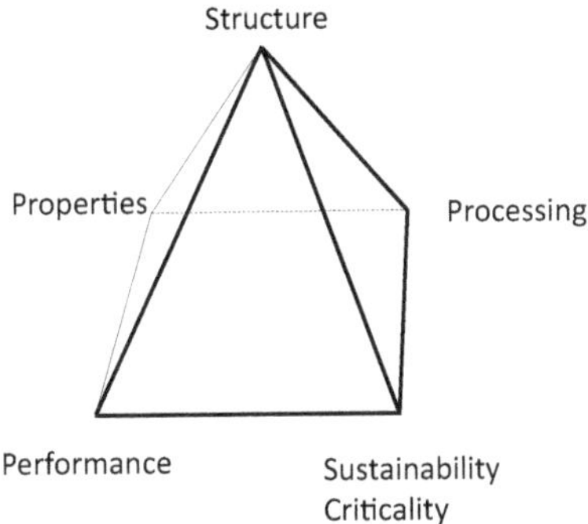

Figure 4.   Materials square pyramid.

Source: Adapted from C. J. Donahue [15].

tetrahedron is too narrowly focused on one moment of the life of materials. It overlooks that materials have a life cycle: where they come from and how they will be disposed or recycled are crucial variables that have to be taken into account as well. Due to the increasing importance of environmental and safety issues in most industrialized countries, many puzzles arise at the end of the value chain. Therefore, a fifth parameter should be added. It has been named sustainability and criticality [15]. Sustainability is incorporated to ensure the life cycle is taken into account and criticality, to address issues of supply and demand, of market and geopolitics. The tetrahedron is consequently reimagined into a pentahedron, i.e. a square pyramid (Figure 4).

Thus, the emergence of the generic concept encompassing a plurality of materials defined as multifaceted material bodies resulted as much from a smooth disciplinary evolution of handicrafts tradition toward science-based technologies as from military pressures in the context of the Cold War and later from a fierce economic competition increasingly constrained by environmental and health issues.

3. Why a Biographical Approach?

We have seen that moving beyond the multiplicity of individual materials was the key to "modern science", and that materials scientists have been able to discipline the jungle of materials thanks to a network of concepts embracing physical features and sociotechnical aspects. It thus may seem rather inappropriate to write biographical narratives of individual materials. Scientists and historians may agree on writing "monographs" of materials, but they are usually more reluctant to use the biographical genre. The Greek term *bios* (translated as life) refers to a human mode of life or manner of living. The extension of the biographical approach to materials not only blurs the boundary between the living and the non-living, but it also overlooks the Aristotelian distinction between *zoe* (referring to the general phenomenon of life characterized by the dual condition of generation and corruption or life and death) and *bios* (the life of moral and political individuals). The use of the term "biography" for stories of non-humans can consequently be criticized. In particular, Thomas Söderqvist, a historian of science and author of scientific biographies, objected that it is non-sense, a marketing gambit for selling books [16].

Indeed, the term "biography" is partly used here in a metaphorical sense, as I do not pretend that materials are behaving like conscious persons. This volume does not play with the rhetorical devices used by popular science writers who personify elements. Sam Kean for instance writes about « aggressive oxygen (which) can dictate its own terms and bully other atoms » while « poor friendly carbon » has « low standards for forming bonds » because it needs four additional electrons on its first shell to make eight and comply with the octet rule [17, p. 5, 34–35). Anthropomorphizing objects may be useful for didactic purposes, but the choice of the title *Biographies of Materials* for this volume is driven by other reasons. Biography as a genre is fully adequate and legitimate for various reasons.

First, materials have a lifespan. They come into existence in certain conditions. Their stories began billions of years ago in the stars when cosmic explosions generated iron, carbon and other elements charted in the periodic table. Of course, they did not exist as materials, i.e. as hybrid entities of nature and technology, before the emergence of humans. They first existed in the cosmos, in the stars, then on the Earth. They had a long and quiet life in the rocks before the apparition of life, when they entered in the more turbulent biosphere. It is precisely the contrast between their long period of existence in nature and their shorter life entangled with human history that legitimizes the biographical approach. Historian of science Lorraine Daston argued that writing biographies of scientific objects is a way to reconcile reality and history [18]. Historicity should by no means be limited to human agents. Objects such as materials or microbes do have a history. Object biographies reconcile two antagonistic views underlying the practices of the history of science. On the one hand, the standard history of scientific discovery relies on the realistic assumption that scientific objects pre-existed their discovery and treatment by scientists: oxygen was already inhaled and consumed by animal respiration before Lavoisier named it, and microbes pre-existed Pasteur. On the other hand, for constructivist historians, scientific objects are invented by scientists and come into existence when they are isolated, characterized and named. A number of chapters in this volume — clays, aluminum, nanocarbons and lithium among others — emphasize the amazing diversity of the modes of existence of materials.

In addition to bridging the gap between realism and historical relativism, the biographical narrative is perfectly adequate for a history of materials because they are neither immutable nor stable. Most materials — including diamond, an icon of durability and a symbol of eternity — have a life cycle. They come into being and they disappear. Materials are ageing and the degradation of their structures is a major concern for the safety of infrastructures. Moreover, the biographical approach to materials is useful for raising citizens' concerns about the environmental impacts of their consumerist behaviors. Life-cycle analysis often shows the tremendous environmental and social impacts of the extraction of raw materials and their disposal reveals the gap between the brief lifespans of commodities and the long lifetimes of the materials they are made of. The chapters on cement, plastics and polonium in this volume provide examples of the environmental, political and ethical issues raised by the time gap between the life of materials and the life of technological products.

A second reason in favor of the biographical approach is that like most humans, materials have strong individualities, sharing a number of features with a family or class of materials. Materials' textbooks distinguish four or five classes of materials — metals, polymers, ceramics, composites and semiconductors. However, every material within those classes offers a special combination of distinctive properties and behaviors that make up its identity The idiosyncratic profile of materials is determined not only by their physical and chemical properties — conductor, insulator, combustible… — but by their technological and economic potential as well. A number of biographies in this volume show how conventional materials known since antiquity have been reinvented as new materials and started a second or third life. Traditional materials such as clays, glass and iron have been used and improved in workshops, factories and alchemical laboratories, while their chemical nature was radically uncertain. But the access to their inner structure and composition opens up new opportunities. The identities of materials are not ontologically fixed by their physical and chemical properties. They are forged through contingent historic events and affordances. They remain open even after being stabilized in a technological trajectory. Biographical narratives are especially suited for embracing such contingent processes because they ignore disciplinary boundaries. To tell the life story of a person like Marie Curie, for instance, a biographer has to go through the history of physics and chemistry, of medicine and institutions, the national history of Poland, the history of women in science…. Similarly, biographers of materials have to combine the history of physics, chemistry and engineering with social, economic, cultural and geopolitical histories along with environmental sciences and regulations. Materials invite to write "compound histories" [19].

Third, materials are not just inert matter, they are active and reactive. As agencies they are put to work by human actors who extract them, process them, use them and rely on them in their technological projects. To fully recognize the agency of materials, we have to move beyond the conventional paradigm of technology inherited from Aristotle. He argued that creating an artifact consisted in bringing together matter (*hyle*) and form (*morphe*). Form is imposed from outside by a human agent onto a passive matter. This hylemorphic model of the creation of artifacts prevailed for centuries in Western culture as an echo to the great divide between humans and nature, between subject and object established by modern science and philosophy. Agency was placed on the side of human subjects, while matter was deprived of activity by modern physics.

However, looking more closely at artistic and engineering practices, the divide between subjects and objects become unclear because materials are never entirely passive. This is not to deny the role of the designer's intentionality, but to realize that between designers and materials, there is no such thing as a one-way relation. Materials are constantly responding and interacting. They have the power to make a difference, to influence the result and affect the designer's project. The dialogue between the material's affordance and the designer's project is precisely the leverage point of creativity.

The agency has to be distributed between humans and materials. Without claiming that "all matter is pulsing with life" like political theorist Jane Bennett, [20] it is important to recognize that materials are connected with their environment. Like all chemical bodies materials are context-sensitive, and their behaviors always depend on the physical and chemical environments. The strong connections established between humans and materials along history should not conceal the existence of relations between materials and forces. Throughout its life, every material is reacting to various forces around: to cosmic forces, geological pressures, biological processes, climatic events as much as to humans' pressures.

This is why it is difficult to strictly predict the safety of materials. The chapters on chlorine, white lead, asbestos and polonium in this volume instantiate the potential harms caused by materials. These cases point to the distinction between hazards, which are due to the intrinsic properties of a chemical agent, and risks, which depend on exposure. While materials scientists and engineers can provide evidence that some materials are hazardous and will cause harm because of their structures and properties, the probability of risks has to be assessed according the conditions of use and exposure in the real world. In particular, since most people are exposed to a variety of materials, the risk assessment should anticipate potential cocktail effects. Like humans, materials are embedded in a dense network of interactions between heterogeneous actors that affects their behaviors. Although it is impossible to give a quick description of these interactions in a few thousand words, biographical sketches are apt to convey a view of the complex and open-ended processes of transformation where materials are engaged.

4. How to Organize a Collection of Biographies?

You will not find in this volume every story that is worth knowing about materials. Even a big atlas would not cover all materials that have enabled technologies and shaped human civilizations. Selecting a few exemplar materials was the obvious solution.

But how to select the samples? In a standard course of Materials Science and Engineering you would pick up one or two materials representative of the five classes of materials — metals, ceramics, polymers, composites and semiconductors. But this conventional classification does not provide any clue for a historical perspective on materials. While it distinguishes semiconductors, useful in recent technologies, it does not include glasses, textiles or pigments, all classes of materials that were extremely important in human history. The conventional taxonomy of materials itself is a historical construction combining heterogeneous criteria such as structure, composition, properties, that requires a historical enquiry.

Since one major purpose of this volume is to open the readers' eyes on the central role of materials in human societies, it could not be organized along the typology of textbooks devoted to purely scientific and technical aspects of materials. Considering materials as hybrid entities, interweaving nature, technology and culture, this volume tries to convey a sense of the social life of materials, and the technological adventures they are part of. Therefore, it gathers contributions by material scientists and by historians of science and technology. Although some chapters are more technical than others, depending on the backgrounds of their authors, all of them insist on the historical roles and careers of materials. The volume is comprehensive in the sense that it seeks to display the widest spectrum of technological, political and environmental issues raised by materials.

The first section provides samples of the most ordinary materials that we use in everyday life. Clay, glass, steel and cement challenge the common distinction between old and new materials. They have been crafted and used by humans since ancient times and they are still objects of intense research and innovations in the 21st century.

Section 2 features materials with a higher profile, because they epitomize the modern ways of life. Some of them like aluminum or plastics are highly visible in daily life, others like silicon lie in the blackbox of electronic devices but they have been driving the pace of innovations in information and energy technologies over the past decades. The life of nanocarbons is even less visible because these new allotropes of carbon are still in their infancy, between laboratory and industry. As they haven't yet been mass-produced, it is too early to describe their achievements. However, the four biographical stories of this section nicely exemplify the integration of materials in technological systems. Innovations do not work in isolation. They also invite reflections on waste that become more prominent in the next section.

Section 3 is no longer about success stories. It features materials that started brilliant industrial careers and became extremely controversial. Chlorine, white lead and asbestos have been mass-produced and used in a variety of applications until they proved to have disastrous impact on human health and the environment. Plutonium, by contrast, is the product of a unique technological sector, strictly regulated. The controversial materials of this section invite discussions of risk issues, secrecy, social ignorance and (ir)responsible technologies.

The final section (Section 4) provides samples of critical materials, materials that generate a situation of crisis because they become vital for warfare or welfare in specific historical circumstances. The three biographies of rubber, lithium and rare earths provide support for better understanding the economy and geopolitics of materials. The criticality of materials is first and foremost a supply chain issue. The scarcity of raw materials at the beginning of the chain creates a bottleneck that paralyzes the production of indispensable commodities. But materials can generate crisis for other reasons such as their impact on the environment or on international relations.

The stories of materials in this volume can be read separately, or section by section, for curiosity or teaching purposes. However, altogether the chapters in this volume converge to shape the Janus face of materials. They have a dual profile as affordances and as components of the world. Their behavior and performances are heavily influenced by what they afford to human actors. The term affordance, introduced by James J. Gibson, does not mirror the intrinsic properties or dispositions of individual substances, but the performance, service and experience that they deliver to its user in definite circumstances [21]. As affordances materials are actors-specific and system-related. Their operation as socioeconomic agencies in technological systems should not blind us about their other face as world components. Whether they are extracted or man-made, materials are objective agencies that have a causal power in ecosystems. They have a life of their own. They have individual actions on, and reactions to, their environment. Hopefully, this collection of stories illustrating the double life of all materials will help materials science textbooks extend the well-established "materials thinking" to a more holistic approach including "ecosystem thinking".

References

1. Smith Cyril, S. (1981). *A Search for Structure — Selected Essays on Science, Art, And History.* Cambridge: The MIT Press. Manchester: Manchester University Press

2. Ashton, T. S. (1951). *Iron and Steel in the Industrial Revolution.* 2nd Ed.; Terence, B. A short history of steel and the Bessmer process, Manchester: Manchester University Press, https://www.thoughtco.com/a-short-history-of-steel-part-ii-2340103; Meikle, J. L. (1995). *American Plastic: A Cultural History.* New Brunswick, NJ: Rutgers University Press.

3. Rademaker, M. (2019). The geopolitics of materials: How population growth, economic developement and changing consumption patternes fuel geopolitics. Chapter 2 in *Critical Materials.* E. Offerman (Ed.). World Scientific Series in Current Energy Issues. Vol. 5. https://www.hcss.nl/news/geopolitics-materials-now-available-online.

4. Geiser, K. (2001). *Materials Matter. Toward a Sustainable Material Policy.* Cambridge: MIT Press.

5. National Science and Technology Council. (1999). *Nanotechnology: Shaping the World Atom by Atom.* Washington, http://www.nanowerk.com/nanotechnology/reports/reportpdf/report23.pdf; Berube, D. M. (2005). *Nano-Hype. The Truth Behind the Technobuzz.* New York: Prometheus Books.

6. Johnson, A. C., Jin, X., Nakada N. and Sumpter, J. (2020). Learning from the past and considering the future of chemicals in the environment. *Science.* 367, 384–387.

7. Krausmann, F., Lauk, C., Haas, W. and Wiedenhofer, D. (2018). From resource extraction to outflows of wastes and emissions: The socioeconomic

metabolism of the global economy, 1900–2015. *Global Environmental Change.* https://www.sciencedirect.com/science/article/pii/S0959378017313031.

8. Mauss, M. (1945). Conceptions qui ont précédé la notion de matière. In *Qu'est-ce que la matière? Histoire du concept et conceptions actuelles.* Paris: Presses universitaires de France.

9. Bachelard, G. (1938). *La formation de l'esprit scientifique.* Paris: Vrin.

10. Venel, G. F. (1751–1765). Entry "Principes". In Denis, D. and D'Alembert Jean Le Rond (Eds.) *Encyclopédie ou Dictionnaire raisonné des sciences, des arts et des métiers.* Vol. 13. https://artflsrv03.uchicago.edu/philologic4/encyclopedie1117/navigate/13/1487/?byte=3963923.

11. Galileo, G. (1638). *Two New Sciences.* https://oll.libertyfund.org/title/galilei-dialogues-concerning-two-new-sciences.

12. Guillerme, A. (1994). *Bâtir la ville. Révolutions industrielles dans les matériaux de construction, France, Grande Bretagne (1760–1840).* Paris: Champs-Vallon.

13. Hoddeson, L., Braun, E., Teichman, J. and Weart, D. (Eds.) (1992). *Out of the Crystal Maze. Chapters from the History of Solid State Physics.* Oxford/New York: Oxford University Press; Cahn, R. W. (2001). *The Coming of Materials Science.* London: Pergamon.

14. Bensaude-Vincent, B. (2001). The construction of a discipline: materials science in the USA. *Historical Studies in the Physical and Biological Sciences.* 31(Part 2), 223–248.

15. Donahue, C. J. (2019). Reimagining the materials tetrahedron. *Journal of Chemical Education.* 96, 2682–2689.

16. Söderqvist, T. (Ed.) (2007). *The History and Poetics of Scientific Biography, Science, Technology and Culture.* Aldershot and Burlington, VT: Ashgate.

17. Kean, S. (2010). *The Disappearing Spoon and Other True Tales of Madness, Love, and the History of the World from the Periodic Table of Elements.* New York: Back Bay Books.

18. Daston, L. (Ed.) (2000). *Biographies of Scientific Objects.* Chicago: Chicago University Press.

19. Roberts, L. and Werrett, S. (Eds,) (2018). *Compound Histories. Materials, Governance and Production.* Leiden: Boston Brill.

20. Bennett, J. (2010). *Vibrant Matter. A Political Ecology of Things.* Durham: Duke University Press. Ingold, T. (2013). *Making: Anthropology, Archaeology, Art and Architecture.* London: Routledge.

21. Gibson, J. W. (1979). *The Ecological Approach to Visual Perception.* London, NY: Routledge.

Part 1

Old and New Materials

Bernadette Bensaude-Vincent

The materials portrayed in this part are champions in terms of longevity: they have been crafted in Antiquity and they are still massively used today. Moreover, these traditional industrial sectors are not lagging behind in routine production. They still attract a lot of Research and Development (R&D) and prove to be still innovative. Clays have always been used by prehistoric humans and they can nevertheless be manufactured today as high-tech nanomaterials. Glass, an ancient material widely used for making windowpanes, acquired a new identity as glass fiber in information technology in the 20th century.

However, the technological innovations in the production processes do not mark the end of the manufacture of conventional materials for ordinary applications. Technical clays, glass, steel and concrete do not overtake the production of standard ones. As high-tech materials are added to the standard clay, glass, concrete and steel, there is no transition from old to new materials. These mundane materials challenge the distinction between old and new materials as they are continuously rejuvenated through intensive R&D. Steel, an alloy of iron, is both a mass-produced commodity with a huge geostrategic impact and a "subtil", a highly sophisticated material. Its identity depends on the process of making as much as on its chemical composition.

Globally the biographical narratives in this part highlight the respective roles of artisan empirical practices and of scientific knowledge. They invite general reflections about the relations between *making and knowing*. Clays have been used for millennia for modeling pottery and house building purposes, while their inner structure remained unknown. Similarly, the processes to cast and blow glass have been developed and improved without access to the inner structure of glass. High quality steels have been produced long before scientists developed a theoretical understanding of what happens during the process of melting, casting and quenching. One major input of the access to the microstructure of old materials is that they open up a class or family of materials. For instance, clay became a member in a big family of porous materials at the cutting-edge of research in nanochemistry. Glass is a generic name for all disordered, non-crystalline microstructures.

In addition to scientific advances, societal or economic demands may also afford new careers to old materials and prompt a new mode of existence for them. Glass used for making glazing, bottles and windowpanes during previous centuries entered into new adventures in its fiber mode of existence for communication technologies. Clays have been reinvented by imaging techniques as porous materials with remarkable sorption and catalytic capacities in the mid-20th century.

Beyond the common features of these life stories, they also display striking contrasts. Clay distinguishes itself in this part with three lives: a natural history counted in millions of years because it's part of the Earth's history, a social history dating back to prehistory and finally a scientific life as a research tool in nanochemistry. Glass by contrast is not abundant in nature, it is essentially a man-made material. Cement and steel are only manufactured products although they retain strong links to nature since the varieties of steels and cements have long been distinguished by their geographic origins, like minerals, and also because their process of production has strong impacts on the environment.

Finally, clays, steel and glass are materials of many uses. Like a Swiss army knife, clay is useful in a wide range of applications: building, agriculture, pottery, medicine, scientific research and cosmetics. Glass and steel are also multifaceted materials that can be used in a variety of sectors ranging from fine arts to heavy industry via information technology. In contrast, concrete is a material designed for one specific function: construction. In this respect, it played a key role in the emergence of civilizations as it fostered the urbanization of human populations. Over the centuries it catalyzed and reshaped the modes of construction through its alliance with steel. However, this wonder material of modern urban life turns out to be hardly sustainable. Not only does it cause severe damages to the environment, but as urban construction is encouraged by land speculation, people tend to save in concrete. In some countries like Senegal, for instance, many people get tied up with unfinished buildings for lack of mortgages.[1] A lot of knowledge and know-how are needed to maintain the life of cement as the master building material.

---

[1] *The Economist*. (May 1, 2021). "A room without a roof".

Chapter 1

Clays and Zeolites

Frédéric Thibault-Starzyk

CNRS. Maison Française d'Oxford, UK

1. Introduction

Clays were among the first materials used by mankind. Raw clay has probably always been used for modeling small figures of animals or humans and for building houses. The discovery and control of firing opened the way to the production of dishware and provided a long-lasting and resistant building material. The improvements of analytical tools in chemistry during the 20th century brought about a better understanding of the particular structure of porous metalosilicates, and with it the purpose design of chemical objects where materials properties can be tuned to control reactivity in industrial reactors or to explore elementary reaction mechanisms.

2. Minerals with a Geological Life Cycle

Clays and zeolites have accompanied humankind all along its history, but they are also parts of the Earth's history. As minerals, clays and zeolites appear, live and die in a cycle governed by tectonic movements in the Earth's mantle [1,2]. Deep under the surface, under very high pressure and temperature, rocks are formed. They are then brought to the surface where they experience completely different conditions, not ideal for them and under which they are not completely stable. When in contact with water and atmosphere, these materials undergo deep transformations. The rocky siliceous substrate is altered while it progressively emerges in an oxidizing environment, under low temperature and pressure. In a wet medium, chemical alteration transforms the substrate at its core into secondary minerals. Depending on the acidity and chemical composition of the surrounding aqueous mixture, some components of these minerals will be drawn away and lixiviated. The contents in silicon, aluminum and other metals will change, leading to different families of minerals.

The results form the basis of clays, with very small particles (often below 2μ), with invisible crystals, often lamellar. Surface erosion will draw these fine powders to sedimentary deposits.

These clays can then be buried under more recent deposits, and progressively go back to deeper layers. With increasing pressure, a partial metamorphism can take place (diagenesis) to form mudstones, shales... At greater depths, with tectonic forces, theses rocks can in turn be metamorphized into slate or into granites, which will eventually come back to the surface, where they will go through hydrothermal alteration and start the cycle again.

Clays are thus mainly aluminosilicates at very fine granulometry. Their crystal structure is mostly in layers. In the presence of water, these fine particles and these layers will give clays their plasticity. The interlayer space can vary with the water content and with the size of cations located between the layers.

3. Ubiquitous Materials

Clays are of utmost economic importance at the world scale because of their many uses and applications. They can be found in all aspects of our everyday life, from medicine to cosmetics, from paper to dishware, and of course in buildings and agriculture. It would not be possible to overestimate their usage and significance.

The definition of clays is based on their natural origin and on their plastic properties when wet. Most of clays' applications come from their plasticity. The structure of clays remained a mystery during most of their history: the particles of which they are formed are invisible to the naked eye, their crystalline structure was only made clear with modern characterization techniques. They are also complex mixtures.

Clays are mixtures of clay minerals together with coarser particles such as sand, with various compositions. Some other aluminosilicates like zeolites can also be found. During the chemical alteration of rocks or during the diagenesis of clays, hydrothermal conditions can lead to the crystallization of new solid phases, not in layers but in a framework of regular channels and cavities: zeolites. Clays and zeolites are thus parent materials, formed by similar processes but under different conditions. Clays are the largest part of the aluminosilicates' family, but zeolites have a unique place, even more for the chemist than for the mineralogist, as we will see later on.

Zeolites often form magnificent macroscopic crystals (Figure 1). They were first described by the Swedish mineralogist Axel Cronstedt in the middle of the 18th century. Confronted by this new mineral, as would any analytical chemist of the time, Cronstedt fired it on the burner, and found an unexpected behavior: the rock seemingly boiled. He then coined

Figure 1. A natural sample of heulandite and stilbite zeolites (20 cm in its largest dimension). Personal collection of the author.

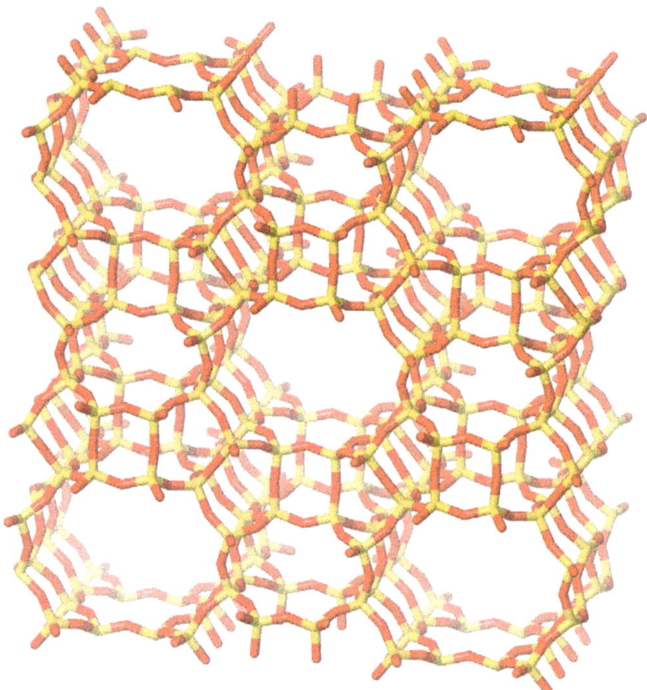

Figure 2. Schematic structure of the mordenite zeolite. The O atoms (in red) link together the Si or Al atoms (in yellow) forming a framework of tetrahedra. The large channels are visible through this slab of the crystal, they are 7 Å in their largest width. Copyright © 2017 Structure Commission of the International Zeolite Association (*IZA-SC*).

a new name from the Greek words for "boiling stone" [3]. With modern techniques, we now know that zeolite are in fact real crystalline sponges, micro (or "nano") porous silicoaluminates.

Zeolites can be considered as a network of molecular-size cavities. They are made by an assembly of silicon and oxygen tetrahedra (Figure 2). Cavities formed in these atomic scaffoldings can range from a few angstroms to 1 or 2 nanometers. These are the exact dimensions for small molecules like water (3 Å). Water, when trapped in a zeolite with the appropriate size, will escape during heating, with the stone itself seemingly boiling.

Cavities in zeolites can accommodate many important molecules for the chemical industry: benzene and other aromatic molecules (6 Å), hydrocarbons… The tight fit between the crystalline host and its molecular guest leads to unique perturbations and new properties, with specific electronic excitations. This is known as chemical confinement and we will come back later to these interactions in nanomaterials.

Zeolites were discovered much later than clays. They were put to great use when large enough sedimentary deposits were found to allow industrial exploitation and when their structures and particular properties were progressively unraveled. Zeolites came to a major place in the chemical industry during the second half of the 20th century, each of their properties affording a new application. Because of their very large sorption capacity they are great materials for absorbing spillages, for use as chemical reservoirs, as food supplement or as soil improvements in agriculture. Capable of exchanging ions with their environment, they are used as anti-scaling agents or for pollution remediation. Zeolites can be acidic or basic, with some very strong sites. They can present multiple sites for chemical reactions and are now largely used in catalysis and petrochemistry.

Initially found as natural rocks, most zeolites have been artificially created for technological or scientific purposes. Contrarily to clays, a wide range of zeolites are thus from synthetic origin. Chemists found ways to tune their structure and composition to suit scientific and industrial needs. These minerals became one of the most significant solid catalysts. Artificial zeolites are used wherever very specific (and wealth-generating) applications are considered. The art of synthesizing zeolites with specific properties is a domain that prompts the imagination of contemporary chemists to invent a wide range of porous mineral architectures.

4. Surface, and Sorption Capacities

Clays and zeolites are for the chemist, and like any other solid, characterized by their specific surface. When a solid is dipped into a gas or a liquid, molecules circulating in the three dimensions will come to the surface and attach to it more or less strongly. The solid will then be able to sorb a certain amount of these molecules. A perfectly flat surface will rapidly be covered. Benjamin Franklin famously showed that a teaspoonful of oil poured on the surface of a pond in London could cover more than 2000 m^2 [4]. On a corrugated solid, the rough surface presents holes and bumps which will increase the sorption capacity. When the solid is divided into smaller grains, the number of molecules sorbed on the surface will even increase. One gram of silica, with its fine particles and its highly divided surface, can offer up to 200 m^2. Porous materials present another adsorption place additional to the external surface; circulating molecules can also sorb inside the pores and cavities. This internal surface can be reached through pores, channels and interlayer space. In some clays and zeolites, each gram of dry solid affords 1000 m^2 of surface with sorption capacity.

It is of course with water in zeolites that the sorption capacity is the most striking. Zeolites have thus been used since the 1950s for the drying of gas, natural gas and coolant gas [5]. The sieving properties of the channels in the solid can even improve the drying effect: large molecules like hydrocarbons cannot penetrate the pore system. Very hydrophilic compounds having large molecules will be dried very efficiently by the simple fact that any water they contain gets into the channels easily and is strongly retained in there. Small (mal-) odorous molecules are also very efficiently sorbed and trapped in the channels and cavities of zeolites [6]. These smell-control capacities make them very popular cat-litters, but also food additives for poultry farming: the solid is mixed with the food as pellets. When ingested, it is quickly loaded with nutrients in the alimentary bolus of the chicken and will only release the nutrients slowly during the digestion, thus increasing their efficiency. The mineral also adsorbs and traps ammonia and other malodorous molecules, leading to less smelly droppings. Added to agricultural soil, these solids can also store water or slowly release fertilizers which will have been added beforehand.

Clays and zeolites are excellent for the remediation of leaks of hydrocarbon or other toxics, which they can absorb quickly before they reach the environment.

In zeolites, molecules can only be adsorbed if they can enter and circulate in the pore system which acts as a molecular sieve. This unique property allows the triage of molecules based on their size or shape. This has been applied in the industrial process for the separation of linear hydrocarbons [7]. Molecular sieving is also applied in the separation and purification of gases by tuning the composition and size of the pores in the zeolite crystals. Oxygen concentrators are loaded with zeolites which will selectively sort oxygen from nitrogen molecules for military jet pilots and for medical respirators.

5. Ions and Exchange Capacity

Specific surface and shape of the internal channels in the crystal structure thus govern a large part of zeolites' properties. Their chemical composition nevertheless also leads to very specific applications. Clays and zeolites are oxides, crystalline assemblies of oxygen atoms (bearing two formal negative charges) with silicon, aluminum or other metals bringing various positive charges. The resulting crystal can be bearing a neutral charge when opposite contributions balance each other. It is very often not the case and any resulting negative charge must be compensated by cations inserted in the crystalline structure, between the layers of the clay or inside the channels and cavities of zeolites.

These compensating cations are mobile and can be exchanged with other cations. Depending on the natural source of the mineral, the cation content will be different and the "terroir" where these materials originate will have a strong influence. These cations are also a tunable parameter for the chemist to get the exact properties she or he is looking for. The size of the cation itself, for example, will change the available space remaining in the pores and will change the molecular sieving properties of the crystal. Clays and zeolites are common cation exchangers; they can transport cations to release them where needed, in exchange for others [8].

The most important use of these cation exchangers (in tons) is for water softening. In washing powders, zeolites previously loaded with sodium will release it in the water and trap the calcium and magnesium ions, thus removing lowly soluble carbonates and softening the water for the laundry. Zeolites used in this process are natural additives with relatively low environmental impact.

Water treatment is not limited to softening. Water in swimming pools or in fish hatcheries needs treatment for the removal of ammonia (ammonium) or other charged molecules which could accumulate up to toxic levels. More generally, many pollution remediation processes use ion exchangers. Nuclear power plants produce radioactive ions which can be efficiently trapped by zeolites [9]. Nuclear fuel waste is sometimes stored at the heart of natural zeolite deposits in the ground to improve containment: any leakage of radioactive ions would be minimized by the zeolite layer. After Fukushima's nuclear accident, zeolites were deposited around the site to reduce leakage into the sea, and polluted water was treated and filtrated using zeolites [10]. Zeolites were used around Chernobyl and given to the population as food additives in pills or sweets (for children) as a treatment to purify the body from radioactive ions [11]. Zeolites are now commonly sold (unspecified and with no indication on the composition) on the internet for detoxification or purification from possible heavy metals.

6. Reactive Sites for Catalysis

By changing the chemical composition of zeolites, reactive sites can be created in their pore system. Molecules entering the pores can then be modified and transformed into products for the chemical industry. The chemical composition is mostly altered by replacing the compensating cations with cations of various degrees of basicity, or even by very strong acidic protons. Metals can also be inserted into zeolites, opening access to the complete landscape of chemical reactivity, from acid–base reactions to oxidations or reductions. Molecules coming in contact with zeolites are sorbed in the pores and reach the active site where they will be modified. Then the modified molecules leave the site and are released in the external volume. The reaction site itself remains unchanged inside the solid catalyst, ready for the modification of a new molecule.

Solids are very important in catalysis because they provide a surface for the reaction to take place. In the three dimensions of a gas or liquid, two molecules (or the molecule and the reaction site) have a fairly low chance of meeting each other to react together. When the circulation of the molecules is limited to a surface (after sorption), chemical contact and reaction is considerably favored. In the pores of a zeolite, the situation is even more specific: the molecules need to diffuse to the reaction site or to the other molecule before they can react, and only reaction products that can actually form inside the available space, and that can travel back to the external volume, can be obtained. Heterogeneous catalysis on zeolites thus offers size and shape selectivity for producing chemicals that would otherwise be very difficult to obtain.

Zeolites became in the 1960s the most important industrial heterogeneous catalysts [12]. They brought a complete revolution to oil refining, allowing the use of a much larger fraction of the crude oil through cracking and isomerization reactions. Some zeolites are even used to transform methanol into gasoline, without the need for crude oil itself [13]. Synthetic zeolites are at the basis of catalysis in modern petrochemistry. The ubiquity of these catalysts and the control they allow also lead to their use in biorefineries to convert the complex chemicals from biological sources [14]. It is usually considered that any industrial synthetic material will have met a zeolite at some point in its life.

By performing chemical reactions in the particular environment of the zeolitic pores, new and unexpected phenomena can be observed. When a guest molecule is adsorbed inside a zeolite without water or any other solvent, the zeolite itself plays the role of the solvent and coordinates with the molecule [15]. The zeolite offers the tridimensional environment around the molecule, it can electronically stabilize it and build up hydrogen bonds. Solvation of some organic molecules (3-hydroxyflavone) was shown to be comparable in a zeolite and in methanol [16]. The zeolite can also ionize salts and stabilize the resulting ions.

Water, as any other protic solvent, limits and levels acid strength by its dissociation equilibrium. In the absence of water, protonation reactions in zeolites are not limited to the pH scale and some very unreactive molecules such as acetonitrile or hydrocarbons can be protonated. Some acidic sites in zeolites are stronger than pure sulfuric acid and are therefore described as superacids [17]. The protonation equilibrium has also been shown to be shifted by temperature [18]. In some molecule–zeolite guest–host pairs, the deformation of the zeolite at high temperature favors the proton transfer and shifts the protonation equilibrium. Acidity is thus not the same at room and at high temperature for this particular acid–base pair. For the guest molecule, the zeolite will seem more acidic at 200°C than at 25°C. This is a strong reminder that it is very important to study catalysts under real reaction conditions rather than in the particular environment (vacuum, low temperature) often needed for characterization techniques.

A most unique aspect of zeolites is the possibility to get molecules inside molecular cages. Such cages can have various electronic or mechanical effects on the confined molecule. That perturbation can completely modify the usual chemical properties. We started to mention acidity, solvation and the influence of temperature, but many other cases have been reported.

These effects are similar to the particular state of matter in nanomaterials. Here, the material might be formed of larger particles, but the nano-voids it contains can provide an interesting mirror-image of nanoparticles.

One of the simplest effects of confinement is the protection it can give to reactive and unstable molecules. Lapis Lazuli is a gemstone with a very particular color, forming one of the most ancient and well-known pigments. Ultramarine blue is caused by free radicals on an assembly of three sulfur atoms $S_3^{-\bullet}$. These very reactive radicals would disappear immediately without the constraints of the zeolite cage. Trapped in the small cavities of sodalite, to which they exactly fit, the radicals are protected and produce this uniquely sourced color (at least before a synthetic equivalent was produced in the 19th century) [19].

Transition metal complexes have also been trapped in zeolites' pores [20]. They were synthesized by the so-called "ship-in-the-bottle" method, where the starting elements are assembled inside the zeolite's internal cavities to form a complex that is too large to escape once it has been formed. The metal complexes can mimic the active center of a natural enzyme. These "Zeozymes" reproduce the chemical behavior of enzymes (natural catalysts), while their mineral matrix protects them against the very harsh conditions needed in a chemical reactor (temperature…). Zeozymes are some sort of hybrid materials between an enzyme and a mineral. Zeolites provide the protective mineral cage but allow access to reactive substrates.

7. Ideal Tools for Chemical Research

Zeolites can be synthesized as perfect crystals, with no extra-framework phase and very few crystalline defects when they are not damaged during the process of synthesis. They thus provide ideal tridimensional chemical spaces with particular constraints for testing a research hypothesis or a computer model simulating the zeolite to be synthesized.

The containment of a molecule in a small space also leads to an "electronic" deformation of the molecule, more exactly a deformation of its molecular orbitals and a change in their energy levels. Electronic densities are modified all along the molecular backbone of the molecule. A new electric dipole can appear and local electronic affinities can change significantly. When a weakly basic molecule like acetonitrile is introduced in the pore system of the mordenite zeolite (in its acidic form), the molecule will be squeezed between the walls (Figure 3) and electrons will accumulate around the nitrogen atoms, which will become more strongly basic. This new electron density will, in its turn, create an electric field and change the electron density in the zeolite walls, increasing their acidity [21]. The acid–base pair becomes much stronger and the proton transfer reaction can take place. This only happens at very specific locations in the pore system, where the fit around the molecule is at its tightest. An empty acidic zeolite, with no water, cations or other molecules in it pores, could be considered as a covalent solid, with very low ionic contribution to its internal bonds. It is by introducing more or less basic molecules that the electronic cloud in the solid is disturbed, ionizing progressively the bonds between the solid and the protons it is bearing and thus creating the Brönsted acidity. The acid strength will increase further with stronger mechanical interaction between the solid and its guest. It is therefore very difficult to determine the acidity of the zeolite on its own. In zeolites, acidity only takes its full meaning within a particular acid–base pair and under particular measurement conditions of temperature and pressure.

The limit in space imposed by the pore walls to the guest molecule can constrain its molecular electron orbitals, destabilizing electronic levels [22,23]. Confinement can thus increase the reactivity of molecules. This gives a further advantage to zeolites as heterogeneous catalysts: nano-voids inside the solid lead to many new interesting properties.

Zeolites have deeply modified our approach to materials. This has led to new *operando* methods for the characterization of catalysts and functioning materials: studying a material *in situ* is not sufficient anymore and a thorough characterization must now be done while the material is operating, under normal conditions [24–26]. Studying a material under the

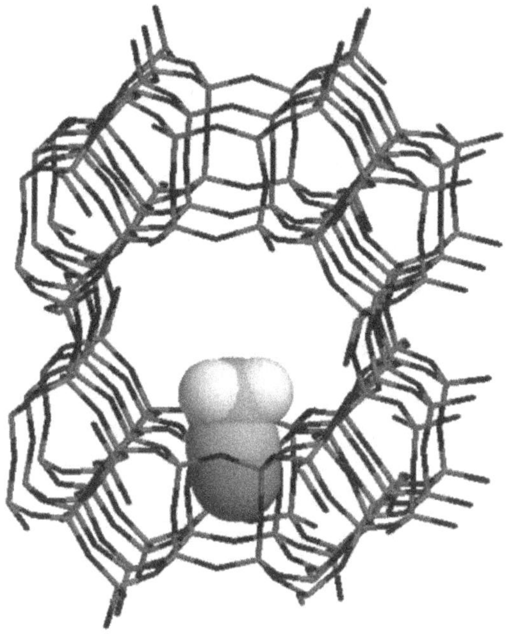

Figure 3.   The molecule of acetonitrile in the pores of mordenite.
Source: Adapted from Ref. [21].

right conditions of pressure and temperature allows a first description of its chemical state, but it is only with the flow of reactants and the changes in the solid during its operation that we can get a detailed understanding of the living material rather than a static and fixed picture. The living world cannot really be fully understood only on a photographic image: similarly, materials can only be investigated when they are in actual operation.

8. Conclusion

Clays and zeolites have been faithful partners of human technology since the dawn of firing techniques and they are still at the cutting-edge of research in the 21st century. Clays have been with us all along our history and they continue to afford a wide range of technological opportunities.

Zeolites were identified more recently, but they opened a new era in our understanding of chemistry on, and inside, solids. We have learned to synthesize these materials and tune them to our needs, so they have eventually become key actors in chemical science and industry. Zeolites shed new light on nanomaterials, allowing the control of chemical matter at the level of atoms. They have changed the chemists' view of surface: when, in a solid, all internal atoms are accessible to external molecules, what is the difference between the volume of the solid and its tridimensional surface? Chemical concepts are easily pushed to their limits inside zeolites. They are a must for advancing research in chemistry, at the boundary between theoretical concepts and the real world.

References

1. Millot, G. (1979). Clay. *Scientific American.* 240(4), 108–119.
2. Reeves, G. M., Sims, I. and Cripps, J. C. (Eds.) (2006). *Clay Materials Used in Construction.* Vol. 21. London: Geological Society; Engineering Geology Special Publications.
3. Colella, C. and Gualtieri, A. F. (2007). Cronstedt's zeolite. *Microporous and Mesoporous Materials.* 105(3), 213–221.
4. Franklin, B., Brownrigg, W. and Farish, XLIV. (1774). Of the stilling of waves by means of oil. Extracted from sundry letters between Benjamin Franklin, LL. D. F. R. S. William Brownrigg, M. D. F. R. S. and the Reverend Mr. Farish. *Philosophical Transactions of the Royal Society of London.* 64, 445–460.
5. Ackley, M. W., Rege, S. U. and Saxena, H. (July 18, 2003). Application of natural zeolites in the purification and separation of gases. *Microporous and Mesoporous Materials.* 61(1), 25–42.
6. Carter, J. D., Corkery, R., Ma, J. and Rohrbaugh, R. H. (2002). Nanozeolites for malodor control. US Patent; US 2002/0142937 A1.
7. IsoSiv Process Operates Commercially. (April 23, 1962). Linde process concentrates normal paraffins in C to C range to more than 95% with 99.8% recovery. *Chemical & Engineering News.* 40(17), 59–63.
8. Dyer, A. (2007). Ion-exchange properties of zeolites and related materials. In Čejka, J., van Bekkum, H., Corma, A., Schüth, F. (Eds.) *Studies in Surface Science and Catalysis.* Elsevier, Vol. 168, pp. 525–53. Introduction to Zeolite Science and Practice.

9. Dyer, A., Hriljac, J., Evans, N., Stokes, I., Rand, P., Kellet, S. *et al.* (December 2018). The use of columns of the zeolite clinoptilolite in the remediation of aqueous nuclear waste streams. *Journal of Radioanalytical and Nuclear Chemistry.* 318(3), 2473–2491.

10. Lehto, J., Koivula, R., Leinonen, H., Tusa, E. and Harjula, R. (April 3, 2019). Removal of radionuclides from Fukushima Daiichi waste effluents. *Separation & Purification Reviews.* 48(2), 122–142.

11. Dyer, A. (2000). Applications of natural zeolites in the treatment of nuclear wastes and fall-out. In Cotter-Howells, J. D., Campbell, L.S., Valsami-Jones,, E. and Batchelder, M. (Ed.), *Environmental Mineralogy — Microbial Interactions, Anthropogenic Influences, Contaminated Land and Waste Management.* Twickenham, The Mineralogical Society of Great Britain and Ireland.

12. Vermeiren, W. and Gilson, J-. P. (August 1, 2009). Impact of zeolites on the petroleum and petrochemical industry. *Topics in Catalysis.* 52(9), 1131–1161.

13. Chang, C. D. and Silvestri, A. J. (May 1, 1977). The conversion of methanol and other O-compounds to hydrocarbons over zeolite catalysts. *Journal of Catalysis.* 47(2), 249–259.

14. Jacobs, P. A., Dusselier, M. and Sels, B. F. (2014). Will zeolite-based catalysis be as relevant in future biorefineries as in crude oil refineries? *Angewandte Chemie International Edition.* 53(33), 8621–8626.

15. Derouane, E. G. and Chang, C. D. (April 1, 2000). Confinement effects in the adsorption of simple bases by zeolites. *Microporous and Mesoporous Materials.* 35–36, 425–433.

16. Person, A. L., Moissette, A., Hureau, M., Cornard, J. P., Moncomble, A., Kokaislova, A. *et al.* (September 21, 2016). Sorption of 3-hydroxyflavone within channel type zeolites: The effect of confinement on copper(II) complexation. *Physical Chemistry Chemical Physics.* 18(37), 26107–26116.

17. Anquetil, R., Saussey, J. and Lavalley, J. C. (1999). Confinement effect on the interaction of hydroxy groups contained in the side pockets of H-mordenite with nitriles; a FT-IR study. *Physical Chemistry Chemical Physics.* 1(4), 555–560.

18. Czyzniewska, J., Chenevarin, S. and Thibault-Starzyk, F. (2002). Infrared evidence for the reversible protonation of acetonitrile at high temperature in mordenite. *Studies in Surface Science and Catalysis.* 142, 335–342.

19. González-Cabrera, M., Arjonilla, P., Domínguez-Vidal, A. and Ayora-Cañada, M. J. (July 1, 2020). Natural or synthetic? Simultaneous Raman/

luminescence hyperspectral microimaging for the fast distinction of ultramarine pigments. *Dyes and Pigments*. 178, 108349.

20. DeVos, D. E., Thibault-Starzyk, F., Knops-Gerrits, P. P., Parton, R. F. and Jacobs, P. A. (1994). A critical overview of the catalytic potential of zeolite supported metal complexes. *Macromolecular Symposia*, 80, 157–184.

21. Smirnov, K. S. and Thibault-Starzyk, F. (1999). Confinement of acetonitrile molecules in mordenite. A computer modeling Study. *Journal of Physical Chemistry B*. 103, 8595–8601.

22. Corma, A., Garcia, H., Sastre, G. and Viruela, P. M. (1997). Activation of molecules in confined spaces – an approach to zeolite-guest supramolecular systems. *Journal of Physical Chemistry B*. 101, 4575–4582.

23. Derouane, E. G. (1990). Confinement effects in sorption and catalysis zeolites. In Barthomeuf, D. *et al.* (Ed.) *Guidelines for Mastering the Properties of Molecular Sieves*, New York: Plenum Press. 225–239.

24. Thibault-Starzyk, F. and Saussey, J. (2004). Infrared spectroscopy: Classical methods. In Weckhuysen, B. M. (Ed.) *In Situ Characterization of Catalysts*. San Diego: American Scientific Publishers. pp. 15–31.

25. Vimont, A., Thibault-Starzyk, F. and Daturi, M. (2010). Analysing and understanding the active site by IR spectroscopy. *Chemical Society Reviews*. 39(12), 4928–4925.

26. El-Roz, M., Bazin, P., Daturi, M. and Thibault-Starzyk, F. (December 6, 2013). Operando Infrared (IR) Coupled to Steady-State Isotopic Transient Kinetic Analysis (SSITKA) for photocatalysis: Reactivity and mechanistic studies. *ACS Catalysis*. 3(12), 2790–2798.

Chapter 2

Glass

Hervé Arribart* and Eric Le Bourhis[†]

*Laboratoire LPEM, Ecole Supérieure de Physique
et Chimie Industrielles de la Ville de Paris, Paris, France
[†]Institut P', UPR 3346, CNRS — Université de Poitiers,
Poitiers, France

1. Introduction

Glass has a very long history [1]. Known for more than 6000 years, it has been continuously improved by man. Many uses have gradually been found, because it is relatively easy to change its composition, which allows to adjust its physical properties, and because the material can be shaped in different forms: sheet, container, fiber… Applications of glass have therefore been developed in a very wide range of sectors, such as:

— Construction: glazing, thermal insulation (glasswool), …
— Automotive and transportation: windshield and side windows, lighting…
— Tableware: bottles and jars, table glass…
— Energy: glass fibers to reinforce wind turbine blade, photovoltaic panels…
— Industry: technical glass fabrics, glass bond for abrasive wheels…
— Optics and electronics: lighting, lasers, optical instruments, flat panel displays…

— Telecommunications: optical glass fibers
— Medicine and biotechnologies: eye glasses, diagnosis equipment…

For most of these applications, the important physical properties to tailor are the mechanical properties and the optical properties. From these viewpoints, glass has two basic characteristics: it is brittle and it is transparent.

The brittleness of glass is inherent to its chemical composition and to its molecular structure. In common language, 'brittle' means 'delicate and easily broken', which refers to a mechanically weak material. But, the scientific definition of brittle is different: a material is brittle if, when subjected to stress, it breaks without significant plastic deformation. This doesn't mean that it should be weak. And indeed, a material can be both brittle and strong. We will see that glass can indeed be made as strong as steel. Making glass stronger has been a constant objective of glass technologists across ages.

The transparency of glass is related to the nature of the chemical bonds between its constitutive atoms. In contrast with metals, there are no free electrons in glass, the reason why glass doesn't reflect light. Light being not reflected, it can only be transmitted or absorbed. From the first objects — which were not transparent — made for decorative purposes by the Mesopotamian and Egyptian glassmakers up to today's optical telecommunication fibers that reach an ultimate optical clearness, a second constant objective of glass technologists has been to make glass more transparent.

Glassmakers had to deal with the complex interplay between structure, properties, processing and performance. Although, we shall highlight breakthroughs in the following sections of the chapter, it should be emphasized that material science and technology was incremental with improvements of the process conditioning the quality of the glass produced. Pushing the process forward was a constant direction for the glassmakers to achieve stronger and more transparent glass.

We will see that the solutions to the challenges of making glass stronger and more transparent have something in common. It is the quest for material's perfection. Defects and impurities are indeed responsible for mechanical weakness and lack of transparency. The history of glass science and technology (developments in the formulation of materials and especially in the associated manufacturing processes) is deeply marked by the quest for a perfect material.[1]

After rapidly reviewing the basics of glass making, strength and transparency (Sections 3–5), we shall present and discuss the most important milestones in the flat glass and fiber glass history (Sections 6–8).

2. The Making of Glass

There are glasses of various compositions, but the simplest and most common glasses, such as those used for making glazing or bottles, are essentially composed of silica (silicium oxide: SiO_2), sodium oxide (Na_2O) and calcium oxide (CaO). The raw materials are sand, which is 100% SiO_2, soda ash (sodium carbonate: Na_2CO_3) and lime (CaO). All are in the solid state at ambient temperature [2]. The obtained glasses are generally known as soda-lime-silica.

The industrial manufacturing process is a continuous process [3].[2] It begins with the raw materials, as powders, being mixed and fed into a furnace where they are heated (1500°C) and fused. In the furnace the raw materials react with each other. A series of chemical reactions beginning in the solid state leads to the formation of the homogeneous liquid glass of composition $SiO_2.xNa_2O.yCaO$, where the numbers x and y depend on the relative amounts of the different raw materials and are typically close to 0.17 and 0.14, respectively.

---

[1] It is nevertheless important to mention that for some applications glassmakers have on the contrary learned to master impurities, defects and heterogeneities in the material. Thus, coloration of glass, for jewels, tableware, cathedral stained glass... is obtained by doping pure glass with transition metal ions. The same is true for the fabrication of laser glass. In another domain of application, glass-ceramics used for example as cook-top in electric cooker are composite materials in which crystalline precipitates are dispersed in a glass matrix.

[2] Glass melting technology — or technology of glass furnaces — has its own long history and is only briefly mentioned here. For a historical review see Ref. [3], covering melting, refining, homogenizing, volatilization, furnace flow and corrosion.

Silica is thus the main component of glass. It is possible to make glass of pure silica, by using sand as the only raw material. In this case, the furnace's function is simply to melt sand; but the melting temperature being higher than 1700°C, this requires the use of a very high temperature furnace, which was not accessible to ancient glassmakers and is still today costly and reserved to some niche applications. The introduction of sodium oxide, which significantly lowers the melting temperature, has been a decisive step in the development of glass.[3] Calcium oxide is important too because it makes glass more resistant to atmospheric water corrosion, avoiding the aging of the material which would cause loss of mechanical strength (see Section 4).

Once the liquid glass is obtained, the second step of the process begins. There are some very different processes, according to the shape one wishes to obtain. To make flat glass, the liquid is spread on a flat horizontal surface [4]. To make fibers, the liquid is spinnered through small holes drilled in a metallic sieve. To make bottles, the liquid is molded in a metal mold [5]. As the temperature decreases, the material becomes progressively more viscous and eventually becomes solid. This happens at the so-called glass transition temperature (about 550°C for soda-lime-silica glass).

3. The Mechanical Strength of Glass

As already stated, the fact that glass is a brittle material doesn't imply that it cannot be made strong. A brittle material is characterized by the fact that it breaks without plastic deformation. An important consequence is that the mechanical strength of a brittle material is not intrinsic. It depends on the presence of defects, because these defects play the role of concentrators of mechanical stress: in the vicinity of defects, the material is submitted to a local mechanical stress which is larger than the stress applied from

---

[3]According to the ancient Roman historian Pliny, Phoenician merchants moored on the river Belus discovered glass accidentally in Syria around 5,000 BC. Pliny describes in book 36 of his *Historia Naturalis* how by leaving cooking pots on blocks of nitrate near their fire, the merchants discovered glass as the blocks melted and mixed with the sand of the beach to form a non-transparent liquid.

the exterior. Glass free from all defects would have high mechanical strength because the chemical bonds in glass are strong, but the presence of defects can limit this strength by several orders of magnitude.[4]

Many kinds of defects can be found in industrial glass. Every kind of inhomogeneity has indeed to be considered as a defect. One can find bubbles, filled by carbon dioxide coming from the reaction of the sodium carbonate raw material, crystalline inclusions, such as undissolved sand grains or impurity in the raw materials, or amorphous inclusions, which are glasses of different composition in the surrounding bulk glass.

In addition to homogeneity defects in the bulk, surface defects, such as flaws or scratches, are very important because it is virtually impossible to avoid their formation when the manufactured glass artifact is handled.

Surface defects are in contact with the atmosphere and environment, and corrosion, mainly by water, is accelerated at surface defects. It contributes to crack extension and severity. Hence, the glass article looses progressively its mechanical strength or say ages [6].

From the mechanical point of view, a perfect glass would therefore be a perfectly homogeneous glass with a perfect surface protected against mechanical and corrosion aggressions.

4. The Transparency of Glass

The nature of the chemical bonds and the absence of free electrons in glass imply that a pure and perfect glass is perfectly transparent. There are two possible causes for a deviation from this situation. The first one is the presence of heterogeneities, because these heterogeneities scatter light. A first condition for glass transparency is therefore material homogeneity. The second cause of non-transparency is the presence of impurities, i.e. extra atoms, that absorb light. The most effective atoms from this stand point are the so-called transition metal ions, such as iron, chromium, cobalt ... The transition metal ions absorb certain wavelengths of light,

---

[4]The theoretical strength of silicate glass is approximately 15×10^9 Pascals. The average strength of a glass bottle is only around 30×10^6 Pascals.

Table 1. Length of transparency of glass as a function of the iron ion concentration (ppm: part per million; ppb: part per billion).

Iron ion concentration	30 %	3 %	0,3 %	300 ppm	30 ppm	3 ppm	0,3 ppm	30 ppb	3 ppb
Length of transparency	1 mm	1 cm	10 cm	1 m	10 m	100 m	1 km	10 km	100 km

varying depending on the metal, leading to the appearance of a color. For example, cobalt ions cause a blue coloration.

Transition metals are used intentionally to produce colored glass, such as stained glass. But, if on the contrary one wishes to produce colorless glass, one must avoid them. The main problem is iron. Because iron is the most abundant element on earth, it is present as an impurity in all mineral raw materials, including sand, and small quantities (a few parts per million) are enough to produce a detectable coloration of glass (think for example of the green color of window glass when observed on the edge, which is due to the presence of approx. 1% iron ions).

From a quantitative point of view, the absorbance of light passing through a glass item is directly proportional to the length of the path of light through the material and to the concentration of iron ions. The absorbance, and hence the transmittance, is therefore constant if the product of the length of the path of light by the iron concentration is kept constant. Table 1 gives the length over which glass is transparent for different values of iron ion concentrations (the criterion that has been chosen is that the item is still considered as transparent if its transmittance is 10%)[5]: when the iron ion concentration is divided by 10, the length of transparency is multiplied by 10.

5. Flat Glass

Flat glass history covers a long period of time and huge progress has been made thanks to considerable insights into science and processing of glass.

---

[5]These figures have to be taken as orders of magnitude, not as precise values.

Figure 1. Blowing of a hot glass bubble using a long blowpipe [7].

The first flat windows were barely transparent while today electronic displays are used in any smartphone communication.

Flat glass use started when some transparency could be achieved. It seems to date back to the Roman era with artifacts found in Pompeii. Many processes were developed and used in parallel. First, flat glasses were presumably obtained by pouring and immediately pressing and stretching the viscous liquid. Translucency only was obtained with cast glass [1]. After glass blowing was invented by the Romans, the same process could be extended to the fabrication of windows (Figure 1). The glass bubble was either (i) further extended to a long cylinder or (ii) opened and spinned [7,8]. Both cylinder and crown processes required knowhow to handle the viscous liquid to its final form when it was frozen. Such operations were devoted to noble men. It is notable that earlier glass composition did not change substantially over time and neither did its viscous properties [2]. In the case of the cylinder process, a second operation was needed when the cylinder ends were removed and the cylinder was cut lengthwise, opened and flattened at elevated temperature (close to the transition temperature).[6] Today, the different processes can be detected thanks to their marks left (bubble shape, thickness variation, tool marks) and in fact, the obtained flat

---

[6]This process is still in use for colored stained glass in Saint-Gobain Saint Just unit.

Figure 2.   Rolling of a molten glass on a cast table [10].

glasses were far from those known today [9]. Sheet glass could not be used for large mirrors or windows with undistorted vision. It was not until 1695 that plate glass was invented in France and flat polished surfaces could be produced. The molten glass was poured on a large metal casting table and rolled by a heavy metal roller handled by two workmen on opposite sides (Figure 2). After cooling, the plate had to be polished by means of rotating disks [7].

Sheet and plate processes were considerably developed and auto-mated in the beginning of the 20th century in the USA by Lubbers (with cylinder blown up to 13 m length), and in France by Bicheroux and Boudin (with plates as large as 30–60 m²) [4,10]. Still, the division between sheet and plate glasses remained because of the different quality (bubble, surface and thickness defects) obtained. So far, a revolution was expected to come when smooth and mirror finish became available with ribbon invention. Instead of blowing a cylinder, inventors attempted to draw the flat glass ribbon directly. Fourcault and Colburn both proposed a vertical drawing although slightly different, while Pilkington brothers came later with a groundbreaking horizontal process where the flat glass ribbon was floated defect-free on a molten tin bath (patent of 1959). The development of the process was very challenging and success was not immediate since glass manufacturers had invested in the automated plate process [11]. Moreover, the equilibrium thickness of the ribbon is about 7 mm, so that thinner and thicker glasses are obtained by extending or contracting the ribbon. This operation leaves marks on the ribbon sides that are to be removed downstream. Nowadays, most flat glass is produced by float units with production reaching 500 tons/day.

Unfortunately, the defect-free ribbon has to be handled once it leaves the float unit and then starts its surface degradation. In the 1980s, a chemical vapor deposition (CVD) unit was introduced in the float unit to deposit coating on the top surface but still the surface in contact with the liquid tin could not be treated (see also Section 8). As stated, the surface flaws degrade the mechanical resistance when they are formed and when the glass panel ages (Section 4). A remedy had to be found. Prince Rupert's drops obtained by dripping a molten glass in water were known to resist a hammer although they exploded when their extremity was damaged [12]. In a similar way flat glass treatments were developed to set compressive stresses at the surface and hence to close the generated flaws. Thermal and chemical tempering allows for this. In the first case, the glass panel has to be reheated above and close to its glass transition temperature and then cooled down so quickly that the shrinkage between the cooled surface and the hotter interior allows for compressive stresses to build in. Instead, ion exchange is used for replacing sodium ion by larger potassium ion so as to obtain a similar result if the composition is enriched with alumina (aluminosilicate glass). The compressive stresses introduced at the glass surface are of the order of 10^8 Pascals and 5×10^8 Pascals, respectively. Interestingly, not only the glass is strengthened but its aging is also stopped [6].

Architectural glass reveals that considerable achievements have been made in glass science and technology [10]. Its brittleness has been mastered so as to take advantage of its transparency and esthetic. Glass beams support and distribute loads in complex glass structures. Full glass walls cover huge buildings and support wind pressures. Noticeably, when tempered glass breaks eventually, it shatters in small and harmless pieces dissipating the energy stored during the tempering operation. For roofs, laminated glasses were invented for security applications. Several flat panels are assembled with intermediate polymer sheets that after a treatment at moderate temperature (below the glass transition) bound the different elements. On breakage, the glass pieces are retained together while perforation of the laminated panel is made difficult. Similar innovations have been used in the transportation field where complex forms can be produced to support load and fit the specific design and aerodynamics. Tools have been adapted and developed to generate shape and surface compressive stresses at the same time. The highest mechanical resistance is obtained with chemically and thermally tempered panels laminated to produce cockpit windows for aircrafts.

Breakthroughs were still to come with the huge demand for flat displays, portable electronic devices and smartphones. Large-display technology required extreme geometric reproducibility during the processing steps carried out at elevated temperature. That means a glass with a glass transition temperature above that of standard soda-lime-silica glass. For electronic devices, lightweight, strong and flexible flat glasses were required. It is also very challenging to produce very thin, hence flexible screens with uniform optical performance and resistance to mechanical shock. In both cases, new production units had to be developed. Microfloats were scaled down by a factor of 10 as compared to standard float unit and their operation temperatures shifted up. Also, fusion draw units developed in the 1960s revealed to be very successful for that new demand. Such unit operates an isopipe from which glass overflows on either sides before fusing downward to form the glass sheet at the root of the isopipe [6]. The new compositions required, namely borosilicate and aluminosilicate glasses, could be further polished for the thinnest applications and chemically tempered for strength [6].

6. Textile Glass Fibers for the Reinforcement of Composites

As compared to flat glass, fiberglass history covers a much shorter period of time, less than a century as it started only when glass processing was mature enough. Flat glass development allowed pushing strength as discussed in the previous section. So far, surface damage could not be eliminated and strength remained far below its theoretical strength.

Then came glass stronger than steel! This has been a reality since the 1930s with the introduction of textile glass fibers, whose main application is the reinforcement of plastics to obtain a composite. The fibers are a few micrometers in diameter and they are dispersed in the material (generally a polymer) to reinforce, bringing stiffness and strength. The strength of these fibers is of the order of 4×10^9 Pascals, not so far from the theoretical strength of glass, and larger than that of steel. The main applications of glass fiber reinforced composites are in automotive (body panels, bumpers...), electronics (circuit boards...), boats and marine (boats construction...), home and furniture (roof sheets, windows, bathtub...). It has become a huge industrial activity, with a global production capacity of 900,000 tons/year.

It has been a long story before we have been able to develop glass with such a high mechanical strength, which means glass with an almost perfect homogeneity. It turns out that homogeneity defects are also detrimental to the optical quality of glass, because they are sources of light scattering. Therefore, almost all glassmakers, those motivated by optical properties, as well as those motivated by mechanical properties, have always worked on the improvement of the glass homogeneity. It is not exaggerated to say that the obsession to obtain a 'good glass' has mobilized most of the development efforts for several millenniums. But this has been a very difficult task, because the sources of heterogeneities are numerous, and the means of characterization of the material have been lacking for a long time. But progress has gradually been made. It is generally considered that the Swiss optician Pierre-Louis Guinand made the most crucial progress. He studied in a very systematical way the various kinds of defects, and he had the idea of mechanically stirring the liquid silicate as long as possible during the cooling process. The result was very positive: most of the defects were eliminated. Today, in industrial glass furnaces, great attention is paid to mixing: furnaces are equipped with several devices (stirrers, bubblers...) that contribute to the elimination of defects.

But, as discussed above making a homogeneous glass is not enough to get strong glass items. It is also necessary that the surface of the material is defect-free. The uniqueness of the manufacturing process of textile glass fibers that is described in what follows is precisely that the fiber surface cannot be damaged because it never touches any other material as it is protected against mechanical and corrosion aggressions by a polymer coating applied immediately after it has been formed.

Ancient Egyptians had made containers of coarse fibers drawn from heat-softened glass, and the French scientist René-Antoine de Réaumur considered the potential of forming fine glass fibers for woven glass articles as early as the 18th century. But it wasn't until the 1930s that the process evolved into commercial-scale manufacturing of continuous fibers. Patent applications filed between 1933 and 1937 by Games Slayter, John Thomas and Dale Kleist, employees of Owens-Illinois Glass Co., record the key developments that step-changed the industry from producing discontinuous-fiber glass wool to making continuous glass fibers with diameters as small as four microns and thousands of feet long.[7] The patents were awarded in 1938, the same year that Owens-Illinois and Corning Glass Works joined to form Owens-Corning Fiberglas Co., which is still today a major glass-fiber producer.

In this process, the liquid and homogenous glass composition is poured from the furnace into heated refractory channels to feed the 'bushing', which is a box made of a platinum-rhodium alloy that has a lot (from several hundreds to several thousands) of small tubular apertures, approximately 1 mm diameter, through which liquid glass tends to flow, forming droplets. At the starting of the production, these droplets are drawn downwards and they take the characteristic conic shape illustrated in Figure 3; the material then quickly cools and solidifies into fiber. Immediately after they are solidified, the glass fibers are coated with a polymeric sizing and the coated fibers are gathered together into strands that go through further processing steps.

There are two important points in relation with mechanical properties:

(1) During the rapid fiber cooling, glass is almost instantly transformed from the liquid state to the solid state. The perfectly smooth surface

---

[7]For this application, alumina Al_2O_3 and boron oxide B_2O_5 are added to the silicate composition to improve the material mechanical properties.

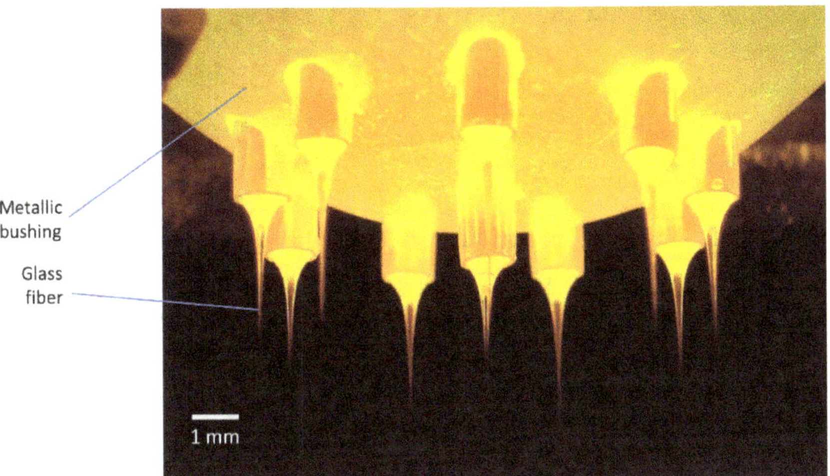

Figure 3.   The formation of continuous glass fibers. The glass is liquid on the top (base of the cone) and solid on the bottom (fiber form). (photograph credit: Johns Manville).

characteristic of the liquid state is thus frozen and the resulting material has no surface defect.

(2) The polymeric coating has several functions, the most important being that it protects the fiber surface against mechanical aggression. For instance, the fibers are not damaged by their contact when the strand is formed. The mechanical strength of the fiber is thus preserved until the fiber is dispersed in the polymer matrix at the time of preparation of the composite in the final application.

7. Optical Glass Fibers for Telecommunications

While transparency had been constantly striven for glass science, the transparency required for communication over long distances was beyond any glass processing for most of the history of glass, as it is highly dependent on the purity of the material. It has been necessary to invent a new way to synthetize the material, in complete departure from the traditional way of making it. Experts in inorganic chemistry, who had followed one another for millennia, had to give the floor to newcomers trained in organic chemistry.

Iron concentration in standard mineral raw materials is of the order of 0.1% to 2%. Table 1 shows that it is not a limiting factor for the transparency of thin items with thickness in the millimeter range. And indeed, ancient glass items in the form of vessels (bottles, bowls…) or sheets have been obtained with good transparency without paying attention to the purity of raw materials (the principal problem to be solved was that of homogeneity defects, as discussed earlier). For vessels, this became common after glass blowing had been invented by the Romans around 30 B.C. For glass sheets, it has taken more time, because there was no satisfactory glass sheet manufacturing process. Romans had initiated the use of glass for architectural purposes. Cast glass windows, albeit with poor optical qualities (due again mainly to homogeneity defects), thus began to appear in the most important buildings of the Roman Empire. But it was not until the 11th century that better glass-making processes were invented. Nevertheless, window glass has long been rare and expensive, and glass has become common in the windows of ordinary homes in Europe only in the 17th century.

Efforts to obtain more transparent glass by selecting purer raw materials have begun in Venice. In the second half of the 15th century, the Venice craftsmen started using quartz sand and potash (instead of soda) made from sea plants to produce particularly pure glass. This opened the way to the development of some new applications of glass, in particular in the field of optics[8] (eye glasses and optical instruments). The first form of eyeglasses was produced in Italy in the late 13th century. The first microscope was created in the Netherlands in the late 16th century, and the first telescope also in the Netherlands in the beginning of the 17th century. As the size of telescopes has rapidly grown, the number of lenses has increased and their size has gotten bigger and bigger. As a result, the length of the total optical path in the glass has increased and reached

---

[8] Apart from the problem of material's purity discussed here, it is important to mention that the development of optical instruments has required the invention of new glass compositions, mainly borosilicates including heavy elements, with better optical properties, and in particular with higher refractive index.

several tens of centimeters. It became therefore necessary to make glass with iron impurities in the ppm range. Selecting the purest raw materials was not sufficient and methods for the purification of the raw materials or of the glass in the liquid state begun to appear. These efforts were amplified during the period of World Wars 1 and 2, driven by the development of military optical instruments. In the mid-20th century, the iron impurity content of the best available glasses was about 10 ppm. They were transparent over distance of the order of 10 meters.

In between, the idea of using glass rods, then glass fibers, to guide light had appeared: when light is introduced at one end of a glass rod/fiber, it travels without attenuation other than that due to the material's imperfections until the other end of the rod/fiber where it escapes, thanks to a phenomenon called total internal reflection. It is thus possible to transmit images at a distance [13]. It is not surprising that the first application of this technique has been in medicine for the inspection of internal organs, a technique known as endoscopy. The first endoscopes had been used during the 19th century. In the beginning, they used rigid glass rods, which was not very comfortable for the patient. It was only by the middle of 20th century that thin glass fibers long enough and transparent enough were available for this application. It then became possible to produce flexible endoscopes.

The development of laser technology was the next important step in the establishment of the industry of fiber optics because this technology had the potential to generate large amounts of light in a spot tiny enough to be useful for fiber optics. The principle of laser has been promoted by Gordon Gould, a Columbia University graduate student who in 1957 joined the private research company Technical Research Group, on the one hand, and by Charles Townes, a consultant for Bell Laboratories, on the other hand. But, the first laser was constructed by Theodore Maiman, a physicist at Hughes Research Laboratories in 1960. Lasers went through several generations including the development of the ruby laser and the helium-neon laser. Semiconductor lasers were first realized in 1962; these lasers are the type most widely used in fiber optics today.

Because of their higher modulation frequency capability, the importance of lasers as a means of carrying information did not go unnoticed by communications engineers. Light has an information-carrying capacity

10,000 times that of the highest radio frequencies being used. However, laser is unsuited for open-air transmission because it is adversely affected by environmental conditions such as rain, snow and smog. Faced with the challenge of finding a transmission medium other than air, Charles Kao and Charles Hockham, working at the Standard Telecommunication Laboratory in England in 1966, published a landmark paper proposing that optical fiber might be a suitable transmission medium if its attenuation could be kept low enough [14]. As a first step, they set the objective of obtaining an impurity concentration of the order of 1 ppm, corresponding to a transparency length between 100 m and 1 km, already enough to distribute information over a campus for example. This article, now considered the foundation of fiber optic telecommunications, had little impact at the time, however.

Charles Kao still had to take an additional step: prove that in some solids the attenuation could actually reach this level of transparency, which he did during the following years by developing an ultra-sensitive optical instrument able to measure very low attenuations with great precision in small samples. He measured in Infrasil®, a vitreous silica from Schott obtained by direct melting of natural quartz,[9] an attenuation lower than this limit. This result was published in 1969 [15]. The two articles, that of 1966 and that of 1969, therefore showed the way: it was necessary to become capable of producing ultra-pure silica in the form of fibers. Charles Kao (see Figure 4) was therefore a visionary. The 2009 Nobel prize in physics has been awarded to him for 'groundbreaking achievements concerning the transmission of light in fibers for optical communication'.

A competition between the Bell Laboratories and the Corning company for the making of the first low-loss optical fiber then started. Bell Laboratories chose to proceed along the lines of conventional glass melting but used specially prepared constituents. Soda-lime-silica and sodium borosilicate glasses were made from materials purified to ppb levels of transition-metal impurities by ion exchange, electrolysis, recrystallization or solvent extraction. But, it turned out to be difficult to avoid contamination

---

[9] Quartz is the crystalline natural form of silicon dioxide SiO_2, silica being its amorphous (or vitreous) form.

Figure 4.   Charles Kao, the 'father of fiber optic communications'.
Source: Wikipedia.

during processing which raises the impurity level from the ppb level in the constituents to ppm levels in the fiber. Corning chose a completely innovative approach, based on the synthesis of ultrapure silica starting not anymore from solid raw materials but from gaseous precursors. This method, known as Chemical Vapor Deposition (CVD), had been invented 35 years earlier by a Corning engineer named James Hyde but had found little application. The advantage of working with gaseous instead of solid precursors is twofold: (i) gases are much easier to purify than solids, and (ii) the resulting material grows without contact with any container that could contaminate it. Adapting this method for the making of a fiber in 1970, Robert Maurer and co-workers Donald Keck and Felix Kapron were the first to obtain a fiber with an impurity concentration less than 1 ppm [16]. Since then, the process has been continuously improved and today the impurity concentration in optical fibers is less than 1 ppb, allowing telecommunication companies to manufacture transoceanic optical fiber cables (Figure 5).[10]

---

[10] In order to compensate for the weak but not zero attenuation, signal repeaters are installed every 100 km or so.

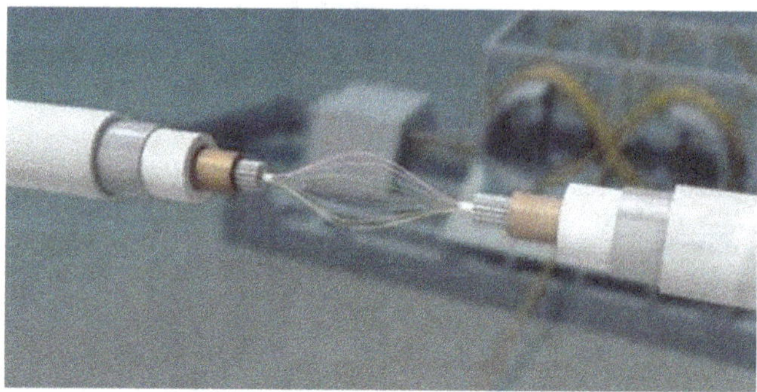

Figure 5.   The transatlantic AEC-1 cable inaugurated in 2016 which comprises eight optical fibers.

Source: Wikipedia.

8. Concluding Remarks

To be used, glass requires strength and transparency. Reviewing glass history illustrates how difficult it was to achieve both. Only when the process became mature enough did the use of glass develop. Important milestones have been highlighted in this historical perspective and most of them are related to the process development. Today's glass processing is the result of many incremental steps as well as a few unexpected revolutions.

Material's perfection has been a long search in the development of glass technologies across ages. Glass perfection was very difficult to reach as it is dependent on the process development. Considerable progress in glass science and processing can be observed from the time when the glass windows were merely translucent. Glassmakers had to play with the mutual influence of structure, properties, processing and performance to improve both transparency and strength.

So far, until recently, remedies had to be applied to neutralize or eliminate defects due to the processing and handling. Only when fiber glass was invented could a perfect glass material be used industrially. Interestingly, for almost all other materials, it is by impurities management, by alloying, by blending, by mixing… that the required properties have been progressively improved and optimized.

References

1. Chopinet, M. H. (2019). The history of glass. In Musgraves, J. D., Hu, J. and Calvez, L. (eds.) *Handbook of Glass*. New York: Springer.
2. For a review of the history of glass compositions, see Kurkjian, C. and Prindle, W. (1998). Perspectives on the history of glass composition. *Journal of the American Ceramic Society*. 81/4, 795–813
3. Cable, M. (1998). A century of developments in glass making research. *Journal of the American Ceramic Society*. 81, 1083.
4. For an historical perspective, see Cable, M. (2004). The development of flat glass manufacturing process. *Transactions of the Newcomen Society*. 74, 19.
5. For an historical perspective, see Dungworth, D. (2012). Three and a half centuries of bottle manufacture. *Industrial Archaeology Review*. 34/1, 37–50.
6. Le Bourhis, E. (2014). *Glass Mechanics and Technology*. New York: Wiley.
7. Diderot, D. and d'Alembert, J. (1765). *L'Encyclopédie*, Art du verre Fabrication des glaces. Paris. https://gallica.bnf.fr/ark:/12148/bpt6k9987c.texteImage.
8. Bontemps, G. (2008). *Glass Making*, Society of Glass Technology. Sheffield, UK.
9. Rasmussen, S. C. (2012). *How Glass Changed the World*. New York: Springer.
10. Hamon, M. (1999). *From Sun to Earth (1665–1999) — A History of Saint-Gobain*. Paris: JC Lattès.
11. Pilkington, L. A. B. (1969). Review lecture: The float glass process. *Proceedings of the Royal Society of London*. 314, 1.
12. Cable, M. (1999). Mechanization of glass manufacture. *Journal of the American Ceramic Society*. 82, 1093.
13. For a history of fiber optics, see Hecht, J. (1999). *City of Light: The Story of Fiber Optics*. Oxford University Press: Oxford.
14. Kao, K. and Hockham, G. (1966). Dielectric-fibre surface waveguides for optical frequencies. *Proceedings of the IEEE*. 113, 1151.
15. Jones, M. and Kao, K. (1969). Spectrophotometric studies of ultra-low loss optical glasses II: Double beam method. *Journal of Scientific Instruments Series 2*. 2, 331.
16. Kapron, F., Keck, D. and Maurer, R. (1970). Radiation losses in glass optical waveguides. *Applied Physics Letters*. 17, 423.

Chapter 3

Steels

Philippe Dillmann* and Catherine Verna[†]

*LAPA, IRAMAT, NIMBE, CEA, CNRS,
Université Paris-Saclay, France
[†]Université Paris 8 Vincennes-Saint-Denis Arscan
UMR 7041 CNRS, France

1. Introduction

Steel has a long history. Until the 19th century, it was a rare, sophisticated material, whose combination of strength and flexibility had been recognized and valued since Antiquity [1]. Steel has fascinated philosophers, alchemists, scholars, merchants and ironmasters because it is both indispensable, and a rather delicate undertaking. Vanoccio Biringuccio (16th century) termed it *subtile*, in homage to its refined qualities.

Today, steel is a common, mass-produced commodity whose production has increased steadily since the 1950s. Output rose from approximately 211 million tons in 1951 [2], of which 95 M were produced by the United States (45%) and 0.9 by China (<1%), to 1.7 billion tons in 2000 [3], including 81 million tons by the United States (5%), 168 million by the European Union (10%), 100 million by the former Soviet Republics (6%), 104 million by Japan (6%), 831 million by China (49%) and 101 million by India (6%). This material is of geostrategic importance because it is vital to industrial development, as evidenced by the increasing role of Asian countries, especially China, in global production over the last 20 years.

Yet, the importance of steel is not new, even though production capacity has exploded in modern times. As early as the 14th century, the Duke of Milan's stranglehold on steel from Brescia, whose production was still irregular, was a key part of this prince's expansionist agenda. Much later, the Cold War magnified the bitter competition between the two blocks that were also fighting via steel. Europe's economic unification began with the creation of the European Coal and Steel Community. Today, the steel giants are China Baowu Steel Group and Arcelormittal. The collective influence that these groups bring to bear on world politics is quite evident.

But when, how and why did steel, which was for so long a rare material, attain this dominance? Is it even the same material? The ways of defining steel have clearly evolved over the centuries; in fact, the modern definition only came into use in the 19th century. Today, steel refers to an alloy of iron and carbon. Compared to iron, the presence of very small quantities of carbon (as little as 0.2%) in steel increases its yield strength and its hardness. Quenching further increases the hardness of steel. However, steel existed long before contemporary science assigned it a calibrated definition based on measuring the amount of one of its components. Since Antiquity, steel has been associated with iron. Until the end of the 18th century, scientists and consumers characterized steel on the basis of its particular qualities. So the question to be examined here is whether the variations in definitions over time are a result of the evolution of theoretical tools, processes, uses and/or markets.

2. Definitions and Designations of Steel

For a long time the Aristotelian definition of steel prevailed: an iron that has undergone various treatments. According to Aristotle, steel is produced by separating the earthy slag from iron by means of successive smelting. Thus, Albert the Great, in his volume *De Mineralibus,* was in full agreement with Aristotle when he described steel as an iron that is refined, pure, hard and very dry; harder and more compact than iron, and which can become brittle with further treatment. This designation of steel as "pure" "refined" iron was fundamental. During the Middle Ages, it was considered common knowledge among the elite, be they aristocrats or merchants. Noticeably the purity of steel resonates with the purification techniques applied to it, constituting the common thread among steel production techniques from the Middle Ages to the present. This idea continued to dominate scholarly circles from the 16th through the 18th centuries, in conjunction with the alchemical tradition. The latter only turned its attention to steel in the 17th century, and then not with the *Acier des Philosophes*, which corresponds to antimony and its derivatives, but with the concept of the *état moyen*. This concept then paved the way for the definition of steel based on its carbon content. It was echoed in the books of secrets appreciated in learned circles during the 18th century, which certainly influenced Benjamin Huntsman (1704–1776), inventor of English crucible steel. For Enlightenment natural philosophers, steel was composed of a metallic base and phlogiston (like all metals) with the addition of carbon. By the end of the 18th century it was clear that iron, steel and cast iron differed by the proportion of their carbon content. It is important to emphasize, in close relation with the scholarly world, the importance of the practical knowledge of craftsmen such as Grignon or Jars in France, and Blakey in England who supported, completed and even critiqued the scientists' theories and experiments related to the question of steel and its definition, especially in dictionaries (E. Chambers' *Cyclopedia,* London 1728; *Encyclopédie* by Diderot and D'Alembert) and technical treatises.

There are various ways of classifying steels. Until the 19th century, steel was often associated with its place of manufacture. Taxonomies also provided buyers with information about the technical conditions during production which, in the eyes of experts and informed consumers alike, were associated with specific areas. In reality, there was not one steel but many steels. In the West during medieval times, the generic term was commonly derived from the term *chalybs/calibs* (which designates both steel and iron). It gradually evolved into *aciare, aciarum* and its derivatives *acer, azero, azale, accai* in vernacular languages. Other terms were often substituted when steel appeared as a particular iron in the case of a heterogeneous product that brings together irons and steels. This was the case for the bloomery reduction in Europe from the Middle Ages until the 19th century (that of the "*molinae/moulines*" and then the *forges à la catalane*) where steel was designated as *fer fort* and *fer cédat*, with differentiated and graduated proportions of carbon, indicating a standardized use of products that were nevertheless heterogeneous.

The names of steels also referred to the trademarks inlaid in the metal indicating the workshop where it was made (for instance, medieval steels from Brescia in Lombardy: *gamba* and *campana* style steels). Buyers immediately recognized these brands and associated the product with the reputation and renowned technical mastery of a given workshop. Most commonly, however, it was the location of production that prevailed, even in the 19th century. An outlier is the emblematic case of "German steel", of which only part was actually imported from Westphalia, Styria and Carinthia. The appellation was, indeed, hijacked on occasion, victim to commercial strategies attempting to take over the market by appropriating the name. But it also made reference to certain assembly techniques of German origin (ex: "English German steel" for English re-carburized steels forge welded into faggots). The modern era was also characterized by practical taxonomies that referred to utilization (razor blade steel, steel for clocks) in order to distinguish specific steels in a market where the competition was escalating, and to ensure appropriate quality for the intended use. Clearly, these designations often combine and restore threads of practical information that can lead to an individual and to his steel. This was the case with Ambrose Crowley's "Crowley steel" near Newcastle (19th century). A final method of designation prevailed in modern times, which led to the identification of processes. This was the case for *acier naturel*, which denoted steel produced by the process of direct reduction and which appeared in the *Encyclopédie*. Yet, the categories can be much more complex. English re-carburized steels provide a good example. These steels benefit from designations that inform the buyer about the repeated operations of cementation and forge welding (successive hammering): shear, double shear, triple shear, among others.... Visual recognition can lead to designations by appearance: cut steel or diamond steel, blister steel, and also identify technical processes.

With the advent of the Bessemer process in the 1850s, the steel industry, which had previously produced predominately heterogeneous mixtures of iron and steel, could finally deliver homogeneous materials. This marked a transition from a world where the location and production techniques made it possible to identify and describe a material and its quality, to a world where the quality itself could be produced at the customer's request. The carbon content of the material and its chemical composition thus became the primary factors in classifying steel. The role of chemical and metallurgical tests became preponderant. With the development of microscopy, metallography and systematic scientific studies at the end of the 19th century, the constituent phases of steel (α and γ iron, perlite, cementite) were identified (notably by Floris Osmond and Henry Le Chatelier in an 1897 article). The earliest versions of the iron-carbon diagram were also proposed (Auzne, 1896), and certain processes, such as quenching, began to be understood experimentally in the laboratory and described theoretically [4]. This was the beginning of a close relationship between laboratory research and industry, which would be strengthened with the production of special steels beginning in the 1880s.

Special steels consist of iron-carbon alloys to which certain quantities of other elements have been added (chrome, nickel, vanadium...) to fine-tune their properties, in particular for the manufacture of weapons. Stainless steels developed throughout the 20th century. The mass-production of special steels had been greatly encouraged by the introduction of electric furnaces in the 1920s. They are particularly well adapted for the production of finely measured grades. Presently, these steels continue to diversify and to address specific needs related to particular uses (corrosion, resistance, hardness, coefficients of expansion...).

3. Early Processes for Steel Making

According to the chronological and geographical contexts, a combination of different disciplines (history, archeology, physico-chemical analysis of the products and waste found onsite and of the objects themselves, situated in their context) is necessary to understand the ways in which steel has been produced.

Steel can be produced directly from the reduction of iron ore and its transformation into metal under the reductive action of the carbon monoxide produced by the combustion of charcoal in a bloomery. Due to the high temperatures attained and the reducing atmosphere, part of the carbon contained in charcoal diffuses into the metal during this operation. So, to a certain extent, this is really a cementation (see what follows) in the reduction furnace, which results in more or less carburizing the bloom. However, the metallic product of the direct reduction remains heterogeneous, with a carbon distribution that can be highly variable depending on the thermodynamic and kinetic conditions inside the bloomery used for the operation. This reduction process was used until the early 20th century in Southern Europe (Iberian Peninsula and the Pyrenees) and even later in Africa. In Japan, after nearly disappearing in the 1920s following the modernization of the country, the process was rediscovered about ten years later (*Tatara*) for making traditional weapons (*Katana*) [5]. In the bloomery, the melting temperature of the metal is never reached, so it is a formless mass (bloom) when removed from the furnace. A certain number of forging and compacting operations (hammering, folding, welding) either by hand or with a hydraulic hammer are required to obtain a semi-finished product (bars, etc.) or a final product. This process is called direct reduction, in contrast to the process for producing pig iron, known as indirect reduction (see what follows). During these post-reduction treatments, the bloom can be treated as a whole. The material keeps the initial heterogeneous distribution of the carbon content (as with the bars from the Roman era found in Mediterranean shipwrecks, or those still in place in medieval cathedrals). The different parts of the bloom (more or less carburized) can also be identified and sorted out by mechanical operations to obtain relatively homogeneous steels from heterogeneous blooms. This delicate operation has been documented in Toulouse (France) in the 14th century by razor manufacturers. Over a very long period of time this technique developed in a wide range of cultures. Comparable practices were attested during the same period in a key steel production area: Styria (Austria). Steel production by fragmentation and sorting was also observed in Africa in the 20th century. Today it is still used in Japan to make *katana* swords (*Tamahagane*).

Another direct method to produce steel consists of deliberately carburizing the bloom, either to make the previously mentioned separation operation highly profitable, or to produce blooms almost entirely made of steel, particularly with the aid of hydraulic bellows. This more systematic process yields larger quantities and may also be linked to the use of manganese ore. In fact, certain regions renowned for their steel production, especially in the Middle Ages (Bergamo Alps, Pyrenees, Carinthia...) are located on iron ores highly charged with manganese.

In Europe and specifically in Rhineland at the end of the 13th century, a new process emerged with the generalized use of hydraulic power: the indirect reduction process presenting several technical variants. The Walloon process, named for its origins in 15th century Wallonia, spread widely in northern France and Great Britain from the end of the Middle Ages and produced pig iron and iron. However, the production of steel with this method remained secondary, at least until the 19th century. Indirect reduction differs from direct reduction in that a higher temperature is obtained in the furnaces (blast furnaces) due to the systematic use of hydraulic power to operate the bellows that introduce air into the furnace. The elevated temperature accelerates the diffusion of carbon from the charcoal (or from coke in recent periods) into the metal, so that its content in the reduction product exceeds 2%, ushering it into the domain of pig iron as defined by contemporary metallurgists (see what follows). As a result of the high carbon content of the metal, its melting point is significantly lower than that of iron; thus cast iron comes out of the reduction furnace in a liquid state. It can then be molded into the form of cannonballs, firebacks, cauldrons, etc. Its high carbon content gives it another property that differentiates it from iron and steel: it becomes brittle under the hammer of the forge. It is indeed another material. Remarkably the production of cast iron appeared in China between the 7th and 5th centuries BC, and the processes of decarburization of this cast iron to transform it into iron or steel were known in the 1st century BC [6]. During this period, China was taking a quite different technological pathway than the rest of the world, especially compared with Europe where cast iron only appeared by the end of the Middle Ages. Considering the indirect process, the transformation into iron or steel occurs during the refining stage, which consists of decarburizing the pig iron in a second hearth. This decarburization results from the introduction of air into the hearth and the interaction of the molten metal with the slag. The product of the refining operation is a paste-like mass of metal, very similar to a bloom. Thus, it is clear that if the operation of decarburizing the pig iron is interrupted before the carbon is completely removed from the metal, the product of the refining process will be a steel, more or less carburized depending on the progress of the operation. It seems that steel produced by decarburization in Europe during the Middle Ages was quite rare. It has been documented in France (Dauphiné then Nivernais, Perche, Champagne) in the 16th century. However, a notable increase in French production did not occur until the beginning of the 19th century, going from 1,000 to 3,500 tons per year between 1800 and 1850. However, this progression seems fairly limited when compared with the production of cementation steel (6,500 tons in 1850, see what follows).

4. Methods and Markets

A new milestone was reached in England in 1784 when Henry Cort patented a process for decarburizing pig iron, puddling. Here, the metal is treated in a reverberatory furnace that heats coal in a fireplace separate from the hearth containing the metal, thereby eliminating all sulfur pollution (remembering that the use of coke in blast furnaces for ore reduction appeared in England in the early 18th century and became widespread in the 1750s). All decarburization processes produce mostly heterogeneous metals containing more or less carbon depending upon the refining conditions. However, the part devoted to the production of steel remained minor (the Eiffel Tower is made of puddled wrought iron, it is not made of steel!).

Since the origins of metallurgy in protohistoric times, another way of obtaining steel consisted of forcing carbon into iron brought to a high temperature through diffusion by exposing it to a hardening agent or cement, a highly carburized material or gas. This process known as cementation takes place over several hours depending on the thickness to be carburized. Thus, there is a distinction between surface cementation intended to harden a limited part of an object and the cementation process giving rise to a material that is entirely steel. The first method makes it possible to treat already-formed objects (blades, arrowheads and files as mentioned by the monk Theophilus in his text *Schedula* around 1120). A simple, easily carried out process, this type of cementation is a practice for perfectioning material. Still widespread in modern times, the technique remains in use today. It is often accompanied by quenching, which hardens the steel after it has been heated to a high temperature then rapidly cooled. For earlier time periods, there is little data about the exact nature of the cements used and their efficacity (roots, bones, leather, guano, etc....). Today, these agents are usually gaseous because this state promotes the exchange of matter.

In contrast, the cementation process has been introduced in 17th century Europe, with steel production in a cementation furnace. In the 18th century, this technical innovation of English origin spread in north-eastern England and especially in Sheffield, a steel-manufacturing city since the Middle Ages. The production of English cementation steel from small iron bars was 1,440 tons/year in 1730 and 2,600 tons/year in 1750. The temperatures in the cementation furnaces exceeded 1100°C. Here, the cement was a charcoal powder. However, the use of bituminous coal is documented as early as the 17th century in Coalbrookdale. The carbon content of the material obtained after one week of cementation could exceed 1%. The heterogeneity of the distribution of the carbon inherent to the diffusion process required subsequent forging, accompanied by material separation or even re-cementation to homogenize the material. These different types of treatments made it possible to produce steels of varying quality (blister, shear — see above) for markets ranging from hardware to surgical instruments and musical instrument strings. Certain metals (Swedish irons) were better suited than others for this operation. In contrast, irons containing phosphorus yielded brittle steel. Cementation steels really opened up the markets from the 18th century onwards, and were incorporated into a wide variety of objects. Cementation steel was polished (English-style polishing) and then combined with other metals, copper for example, to produce all sorts of household goods, toys and jewelry.

It is worth focusing a moment on a specific process for manufacturing steel that enables to reach the liquid state, in small quantities, and thereby control the homogeneity of the products obtained: the so-called "crucible" process. The origin of crucible steel has been widely debated, but it surely dates back to the 3rd century AD, or even to the beginning of the Christian era [8]. The term crucible steel encompasses several technical conditions. Vanoccio Biringuccio in *La Pirotechnia* (16th century), which was revisited by Agricola and other authors in the 17th century, described its production in Europe during the Renaissance. Presumably, in a bath of pig iron and other fluxes, small pieces of barely-worked iron, resulting from the decarburization of pig iron, were plunged into a crucible. The extracted product, after undergoing quenching, was thus appraised as steel. Crucible steel was also made in Central Asia, where *pulad* steel was produced, especially in Iran as early as the medieval period. Archeological sites have also been excavated in Uzbekistan and in Turkmenistan, the oldest of which date from the 8th century AD, where crucibles used for steel production have been found. Two techniques for obtaining crucible steel have been developed in this context. One consists of carburizing iron with an organic cement, sometimes by adding a mineral compound. The other technique is co-fusion, which consists of mixing pig iron and iron in the same crucible to obtain steel at the end of the operation. In India at the beginning of the 19th century, the cementation process (known as the Mysore process) was observed by European travelers [7] particularly at the production sites of Chinnarayanadurga and Devarayanadurga, while co-fusion (called the Hyderabad process) was more widely practiced in the district of Nizamabad, in the former state of Hyderabad (now Telangana). Note that the co-fusion process has also been documented in China in the 17th century but may be even older. In Sri Lanka, only the first technique was used, but it spanned from at least the 6th century to the 19th century. In the 17th century, it was the use of a highly carburized crucible steel of Indian origin, called *wootz,* that after forging made it possible to produce blades with the characteristic appearance and "superplastic" properties known as recrystallization Damascus (not to be confused with the pattern-welded Damascus resulting from welding sheets of iron and steel, as with Merovingian swords). Finally, crucible steel also spread throughout the Islamic world. The production of crucible steel has been confirmed in 12th century Seville, the capital of the Almohad Caliphate. Written sources also document the practice in the 13th century in the Middle East and Egypt [9]. The complexity of this process using small crucibles vastly limited production.

The scale was different for crucible steels produced in Europe from the 18th century onward. Emerging in England, processes intended to produce "molten steel" or cast steel, crucible steel, were invented in Sheffield by Benjamin Huntsman and experienced rapid production growth from 1820 onward. They made it possible to homogenize cementation iron, without re-carburizing or forge welding, by incorporating a smelting step in a refractory crucible. In this manner, fragments of blister steel placed in the crucible with wrought iron and a flux are converted in a few hours into molten steel, cast into ingots. The temperatures in the crucible, placed in a furnace fueled by coking coal, enable the complete melting of the metal. In the 19th century, the emergence of large production units made up for the limited capacity of the crucibles. The ingot molds for casting molten steel were gradually perfected as well as the material of the crucibles themselves in the second half of the 19th century. This process persisted until the Second World War (with special steels — see what follows).

5. Industrial Processes

Patented in 1855 and industrialized in the 1860s, the Bessemer process significantly changed pig iron refining. By blowing air directly into molten pig iron inside a converter, the subsequent elimination of silicon and carbon causes an exothermic reaction to occur elevating the temperature to reach the melting point of iron. Through this process, metals with very different carbon contents are obtained and thus, very easily, steel in large quantities. Nevertheless, phosphorus, an element present in significant quantities in some pig irons, is not eliminated in this process. This important limitation is circumvented by the use of a basic lining inside the converter. By forming a calcium phosphate, this lining permits treatment of phosphorus pig irons and thus significantly increases the range of processes for obtaining metal in the liquid state. This new process, known as Thomas-Gilchrist, appeared around 1880. In parallel (1864), the Martin-Siemens process, another innovation, used additions of scrap metal (enabling their recycling) for refining pig iron and similarly reached the melting temperature of the refined metal. For all of these processes, obtaining the final metal in its liquid state made it possible to eliminate a large number of non-metallic impurities by flotation, thereby producing a homogeneous metal and better controlling its mechanical properties. At this stage, the possibility of rapidly producing large quantities of steel with these processes increased production capacities tenfold. This evolution was also driven by growing demand (railway tracks, weapons, machinery, metal constructions...). Steel became a common commodity. Gradually, this material took the place of iron produced by the old processes (puddling especially). Thus, the production of steel definitively exceeded that of iron in England beginning in 1885, and in France by 1895.

In the 1950s, the Martin-Siemens process dominated the global scene even though, in Northern Europe the Thomas-Gilchrist process accounted for two-thirds of the steel production. The use of pure oxygen for refining (known as "basic oxygen steelmaking") emerged in the 1960s and rapidly spread worldwide. Highly flexible, it can be used to treat all types of pig iron. Simultaneously, the abundance worldwide of scrap iron accompanied the intensified development of the electric furnace, which dates from the beginning of the 20th century and enabled recycling by smelting. It accounted for 20% of global steel production in 1990. The development of continuous casting, which increases the forming efficiency of long semi-finished products, is also progressing. Lastly, in the 1960s, specific techniques during metallurgical operations increased control over the composition of the steel in the ladle. The analysis of steels during the manufacturing phases and the instrumentation of these measurements increases control over the composition and the reproducibility of the different grades sought. Since the 1980s, only two coexisting sectors have remained: plants combining the production of pig iron and oxygen refining and those using electric furnaces and processing scrap metal.

6. Conclusion

From this brief survey of the history of steel it is clear that the steel used around the world, from protohistory to the present, is a multifaceted material. Its identity is closely linked to that of iron and pig iron. As outlined here, the diversity of steels depends as much on technical processes as on the uses for which they are intended. Regardless of the fabrication method, the operating chain of steel manufacturing is complex and nuanced according to market demand.

Another striking feature of this long history is that the production processes come from all over the world. Steel has no homeland, it has been produced and improved in many countries, often in relation to weaponry. From the Middle Ages to the present day, steels have responded to requests and commissions: steels for Damascus or *katana* blades, steels for medieval and Renaissance European armor, scythes from Styria from the Middle Ages to modern times, special steels for aeronautics. Since the 19th century, steel has been produced in large quantities (railway equipment and weapons, among other examples). Yet, regardless of the period and area considered, steel is a refined product whose production remains a geostrategic challenge.

References

1. Dillmann, P., Perez, L. and Verna, C. (2011). *L'acier en Europe avant Bessemer. Histoire et Techniques.* Toulouse: CNRS-Université de Toulouse-Le Mirail. p. 530.
2. Un siècle de développement de la production d'acier. (1957). Supplément au bulletin mensuel d'information de la CECA. Luxembourg.
3. *Steel Statistical Yearbook*, International Iron and Steel Institute, Committee on Economic Studies, Brussels (2001).
4. Chezeau, N. (2012). De Réaumur à la Première Guerre Mondiale: les étapes de la maîtrise de l'acier, l'essor des aciers spéciaux. *Comptes Rendus Chimie* [Internet]. 15(7), 585–594. http://www.sciencedirect.com/science/article/pii/S1631074812000859.
5. Michel, X. (2001). *Etude archéologique des techniques d'acquisition et de transformation du métal dans le cadre de la production des lames de sabre au Japon (IXe-XVIe siècle).* Paris.
6. Lam, W., Chen, J., Chong, J., Lei, X. and Tam, W. L. (2018). An iron production and exchange system at the center of the Western Han Empire: Scientific study of iron products and manufacturing remains from the Taicheng site complex. *Journal of Archaeological Science* [Internet]. 100, 88–101. http://www.sciencedirect.com/science/article/pii/S0305440318305508.
7. Craddock, P. (1998). New light on the production of crucible steel in Asia. *Bulletin of the Metals Museum.* 29, 41–66.
8. Feuerbach, A. (2006). Crucible Damascus steel: A fascination for almost 2,000 years. *JOM* [Internet]. 58(5), 48–50. https://doi.org/10.1007/s11837-006-0023-y.
9. Karlsson, M. (2000 December 30). Iron and steel technology in Hispano-Arabic and Early Castilian Written Sources. *Gladius* [Internet]. 20, 239–250. http://gladius.revistas.csic.es/index.php/gladius/article/view/72.

Chapter 4

Cement and Concrete

Henri Van Damme

*Ecole Supérieure de Physique et Chimie Industrielles
de la Ville de Paris, Paris, France*

1. A Challenging Situation

Concrete, the mix of aggregates — sand and gravel — with water and cement, is in the hot seat. Unloved by the majority and yet ubiquitous, concrete is the backbone of our built environment. On equal footing with silicon (for information and communication technologies) and, so far, fossil fuels (for transportation), it is one of the pillars of our developed societies. More concrete is produced than any other synthetic material on Earth. Twice as much concrete and mortar is used in construction — roughly 35 billion tons per year [1] — as the total of all other industrial building materials including wood [2], steel [3,4], plastic [5] and aluminum [6]. Roads, bridges, tunnels, dams, power plants, ports, airports, dikes and seawalls, waste- and fresh water plants and networks, all these infrastructures rely on the extensive use of concrete, just like schools, hospitals, public and commercial buildings. There seems to be no other currently known material available in the quantities required to meet the global demand for building and infrastructure.

Although the word concrete is a generic term that applies to any granular composite to which cohesion is provided thanks to a binding phase — bitumen, clay, epoxy, lime, etc. — modern concrete relies almost exclusively on the use of Portland cement or its variants. Portland cement or, more

55

exactly ordinary Portland cement (OPC), is the ground form of clinker, the artificial rock obtained by heating a mixture of limestone and clay at high temperature, in a –3/1 ratio around 1450°C. The magic of Portland cement lies in its ability to "set" and to harden by reaction with water, even in an excess of water. This property called "hydraulicity" makes Portland cement fundamentally different from its raw constituents.

Not surprisingly, considering the success of concrete, the global demand for cement has seen an unprecedented growth in the last half-century, especially since the turn of the third millennium (Figure 1). In relative terms, the increase of cement demand in the last 50 years has still been surpassed by that of plastic [7], but in absolute terms cement remains by far the largest industrially manufactured product. There is a wide consensus that this exceptional growth of cement and concrete consumption on the global scale is temporary and due to a handful of actors only among the emerging countries, China in particular [8,9]. But there are also good reasons to consider that a slower but lasting growth will go on for long. Developed countries face the challenge of maintaining and upgrading their extensive but aging transport, power, water and telecommunication networks, whereas many developing countries, in Africa in particular, still dedicate a large fraction of their national income to satisfying basic human development needs like access to water, sanitation, electricity and affordable housing.

The energy transition and the ongoing climate change are probably not going to mitigate the needs. Renewable energy facilities like wind farms require a substantial amount of concrete for their implementation and the rise of the ocean levels will likely trigger the construction of thousands of km of protective dams. Altogether, this is likely to drive a lasting increase of concrete demand for decades to come.

In parallel with its galloping dissemination, concrete has made tremendous technical progress. The late 19th century mixes of aggregates, cement and water, produced on site, had a compressive strength of the order of 20 MPa (Mega Pascals)[1] and they were rammed in a barely wet and rather stiff

---

[1]The compressive strength of concrete is measured by applying an increasing vertical load on a normalized cylindrical sample, up to fracture. The vertical stress (load per unit cross-section area) at rupture has the same units as a pressure: Pascals (1 Pa = 1 Newton/m^2; 1 atmosphere $\cong 10^5$ Pa or 0.1 MPa). A pillar made with a 20 MPa concrete could be as tall as approximately 600 m before collapsing under its own weight.

state either between movable forms to make walls or on a falsework to make arches. Contemporary concrete with a predefined compressive strength between 50 and 200 MPa is now routinely produced in ready mixed plants and delivered on site in a much more "workable" (fluid) state.

The manufacture of Portland cement itself has also made valuable progress in environmental terms, thanks to improvement in kiln efficiencies and heat management, type of fuel and reduction in electricity requirements (mainly for crushing). As a result, the production of OPC generates now on average -840 kg CO_2/kg of clinker, as compared to -1 kg CO_2/kg of clinker a few decades ago [10].

In contrast with this techno-economical evolution, the social picture is less engaging. Alternatively lauded or execrated, concrete is also the most controversial among all building materials. In spite — and perhaps because — of its emblematic role in the development of the modern world, it is crystallizing our expectations, our disappointments and sometimes our hates. Constantly oriented toward modernity and heavily loaded with moral values like honesty, simplicity, functionalism, optimism or communalism by star architects and city planners of the modern movement, concrete has been facing the violence of revolt when people came to compare the promises with the brutality of many urban renewal schemes in the late 20th century and the monotony of contemporary suburban development [11,12]. Widely perceived as dull and repetitive — "in short, a sort of frightening metonymy of the industrial age" [13] — concrete has been so far failing to meet its social promises.

In our increasingly eco-sensitive 21st century, concrete is now also blamed for its contribution to climate change through its carbon emissions, coming almost exclusively from cement production. In spite of continuing efforts to improve the efficiency of the high temperature manufacturing process, the cement industry is still responsible for about a quarter of all industry carbon emissions and about 8% of all emissions. Among all major heavy industrial sectors (chemicals, iron and steel, mining, oil and gas), cement production generates the most emissions per unit of revenue. Unfortunately, this is to a large extent (about 2/3) intimately related to the very chemistry of the process, involving calcination of limestone.

Excessive carbon emissions are not the only environmental complaint held against cement and concrete. The ubiquitous use of concrete is also blamed for a variety of environmental problems like loss of farm land and increased vulnerability to natural hazards (flash floods in particular, due to increased imperviousness of soils), destruction of landscapes, loss of biodiversity, destruction of social link, loss of traditional constructive cultures, or depletion of natural resources, sand in particular [14], due to very limited recycling efforts. All taken together, it is an extraordinarily severe indictment that concrete is facing.

2. Not Only a Material, also a Construction System

Concrete is a granular composite material, but in most situations, it is implemented in an even more composite configuration, in association with steel. Actually, whether reinforced with steel rods ("rebars"), tendons, fibers or a combination of those, or even not reinforced at all, concrete is first of all a construction *system*, in which the material itself is intimately coupled to an implementation and a construction method. It was already so in the early days of the mid-19th century, not much after the discovery of modern Portland cement, when the mixture of aggregates, cement and water was implemented in a barely wet state. Concrete became even more system-like when reinforcement was introduced. The hundreds of patents filed between −1870 and −1905 on the subject and the many companies to which they gave birth were all promoting concrete as a particular construction system, with a distinctive combination of matrix, reinforcement, structural type and construction method [15–17].

The best illustration of the "material-system" concept is probably pre-stressed concrete. Pre-stressed concrete was invented in 1928 [18]. By removing the dichotomy between tension and compression (perfectly pre-stressed concrete is supposed to work exclusively in compression), it was a radical change in the way concrete was used and it led to deep modifications in architectural design.

More recent developments like self-placing concrete (also termed self-compacting, self-consolidating or self-leveling concrete) [19] or ultra-high-performance fiber-reinforced concrete (UHPFRC) [20], follow the same trend. The latter in particular brings construction with concrete in the same structural family as construction with steel. A similar rupture may be expected with 3D-printable concrete, which has the potential to lead to a totally new construction method. However, this goes along with new stringent requirements in terms of rheological behavior and self-adhesion [21].

In spite of this evolution toward more technicality and mastered complexity, the construction sector does not benefit from a much better image than concrete. The world is covered with a wealth of iconic buildings designed by star-architects and with impressive civil engineering works that are true architectural and technical feats. Yet, construction is still considered by many as the low-tech end of the industrial landscape. One reason for this poor image could be the notoriously low average productivity of the construction sector. In terms of value added by construction workers per hour of work, construction is indeed performing below other major economic sectors like agriculture, manufacturing, wholesale and retail, or even mining [22]. Labor-productivity growth in construction has averaged only 1% a year in the past 20 years, compared with 2.8% for the total world economy and 3.6% for the manufacturing industry. While the manufacturing sector has implemented entirely new concepts of flow, modularized and standardized designs, and aggressively automated to increase production, the construction industry is still struggling to disseminate its most promising progress, often curbed by the stringent norms and standards that it has imposed on itself [22].

3. Reimagining Cement, Concrete and Construction

Cement, concrete and the related construction sector are clearly at a crossroads. Given the wide availability of silicates and limestone all over the world and the unsurpassed convenience of use of a liquid stone turning to solid rock at room temperature, Portland-type cement and concrete are likely to remain the construction materials of worldwide choice. The unprecedented need for decent dwellings and new durable infrastructure in the global South and the urgent need for renewing aging infrastructure in the North are adding further favorable outlooks to this already comfortable situation. However, as summarized in the previous sections, the sector is facing tremendous environmental challenges related to unacceptably large carbon emissions and to the depletion of natural resources (fine and coarse aggregates) which were once considered as inexhaustible, not speaking of other social aspects.

Unless radical changes are identified, developed and implemented, cement and concrete will soon find themselves in an unsustainable situation under societal scrutiny. The following is a brief survey of the strategies that are currently considered or already introduced to mitigate the environmental impact of cement and concrete production and use.

4. Making Better Concrete with Less Cement

Concrete is a cohesive-frictional granular material and the art of concrete formulation is primarily the art of packing grains [23]. It is the key for controlling rheological properties in the fresh state [24] and for obtaining optimal mechanical properties after setting. At first order, the packing density is controlled by the width and continuity of the particle size distribution. The wider the particle size distribution, the stronger the final material will be. Ultra high performance concretes — which are also very fluid in the fresh state — have a broad and continuous particle size distribution, especially toward the lower end of the particle size spectrum [25].

Obtaining the optimal granular composition under given local constraints for ingredients' properties and availability is not an easy task and a number of quantitative proportioning methods have been proposed [26]. In real practice, things are not that simple, for several reasons. First, a significant part of the particle size distribution — cement, SCM, fillers — is currently imposed by norms (or "codes"). Second, the design of an optimal formulation for a given application, taking into account the available aggregates and additions, is a highly technical operation, the theoretical basis of which is still under development. Finally, a minimum of cement content well above what could be achieved with an optimized granular formulation may be imposed by regulations. In practice also, many builders add an extra amount of cement in their mortar or concrete, just to be on the "safe side". The general result is that, in most situations, concrete has unnecessarily large cement content. As an example, the scrutiny of large data bases [27–29] shows that the cement content of a –20 MPa concrete may vary from a remarkably low –20 kg/m^3 to a wastefully large –400 kg/m^3!

A possible step toward a more frugal use of cement may be the widespread introduction of digital, possibly artificial intelligence-based, formulation methods, concomitantly with the introduction of performance-based rather than composition-based standards.

5. Making Better Cement with Less Clinker

A simple and effective strategy to reduce the global warming potential of Portland cement-based concrete is to use cement in which a less carbon-intensive fine mineral material is substituted for part of the clinker, hopefully without loss or even improvement of durability and mechanical performance [1,30,31]. It is usual to distinguish reactive substitutes from less reactive or inert substitutes. The former are usually referred to as "supplementary cementitious materials" or SCM, while the latter are usually termed "fillers". Actually, there is no such thing as a totally inert addition. All fine mineral additions interfere to some extent with the hydration process, be it by accelerating or retarding nucleation only.

The major families of SCM are silica fume (SF), fly ashes (FA), granulated blast furnace slag (GBFS), metakaolin (MK) and crushed limestone. SF is a byproduct of silicon and ferrosilicon alloys metallurgy. Similarly, FA is the condensed form of the non-combustible part of coal in the exhaust gases of coal-fired plants. Both SF and FA are spherical particles, of sub-μm size for SF and of larger size, comparable to that of the cement grains (–50μm) for FA. GBFS is the residue of cast iron production from the high temperature melting of iron ores with flux materials (Ca and Mg carbonates) and coke. Finally, MK is the product of kaolin dehydroxylation. SF, FA, BFS and MK are essentially amorphous materials.

In spite of their clinker-diluting effect, SCM are able to maintain the level of durability and mechanical performance of unblended cement, thanks to their composition and their amorphous character (for SF, FA, BFS and MK) which favors reactivity. A particularly promising strategy is the *co*-substitution of clinker with limestone and a source of reactive alumina. Limestone reacts with available alumina to form carbo-aluminate phases which contribute to strength and durability [1,30,32]. Metakaolinite is a suitable source of reactive alumina, but calcined kaolinitic clays with a kaolinite content down to 40% can do the job at a much lower cost. Kaolinite rich clays are widely available in many places, especially in inter-tropical regions. This is the basis for the so-called LC^3 blended cement [33]. Co-substitution allows for much higher degrees of substitution than with either clay or limestone alone, down to clinker contents of around 50%. The development of LC^3-type mixes may well prove to be a decisive step toward sustainability [34].

6. Using Locally Available Materials and Skills

As far as construction is concerned, our world is an extraordinarily diverse showroom of resourceful and frugal answers to local material and environmental constraints [35]. This intrinsically vernacular character of traditional construction materials and techniques — to use with measure what is locally available and recyclable and invent locally-optimized solutions — is contrasting with the evolution of the modern construction sector, privileging a limited, uniform and standardized choice of materials and constructive solutions. This out-of-context character of the present construction sector is but a confession of our inability to deal with contextual diversity.

A topic on which cement and concrete would clearly benefit from adaptation to local conditions is that of SCMs and pozzolanic materials in particular. There is no point to force the use of only one or a very limited number of clinker substitutes on the global scale. While fly ash and blast furnace slag may be good and widely available substitutes in some parts of the world, calcined clays, volcanic ashes or vegetable ashes may be a better choice in other places. Sand itself would benefit from a better ability to use local resources.

It is important to remember that the most widely and locally available eco-friendly construction material is raw (crude, unfired) "earth", or soil. In many respects, raw earth is nothing but a natural concrete, the fine (less than 2μm) clay fraction playing the cohesive and adhesive role that cement hydrates play in normal concrete [36]. Raw earth has been used as a construction material for thousands of years on a worldwide scale and it is still extensively in use today.

Not surprisingly, with the increased awareness of environmental and energetic issues, earth construction is gaining renewed interest and several successful attempts have been made to transfer to earthen construction the technologies used in cement concrete construction, like casting in place, prefabrication or even 3D-printing [36–38].

7. Using Concrete More Efficiently Through Smart Design, Topology Optimization and Architectural Geometry

Nature is full of complex, often hierarchical structures which succeed in optimizing simultaneously mechanical, optical, hydrodynamic, etc. properties and economy of material use. They are a fountain of inspiration for architects and engineers [39,40]. Typical examples are the cellular structure of trabecular bone, the hierarchical structure of nacre, or the ribbed structure of some large leaves, which have been an obvious source of inspiration for Pier Luigi Nervi for his amazingly thin floors and roofs which owe their stiffness to their ribbed structure.

What nature achieved through natural evolution, engineers may try to accomplish through topology optimization, also called generative design. Topology optimization is a mathematical method that optimizes material layout within a given design space, for a given set of loads, boundary conditions and constraints [41]. The design can attain any shape within the design space. Optimizing a form regarding its own weight and force distribution leads generally to pseudo-organically grown structures.

Due to the free forms that naturally emerge from this, the result is often difficult to build or to manufacture. This is where 3D-printing is emerging as a promising solution (see Section 9), but recent developments suggest that transformational progress may also come from theoretical research in so-called architectural geometry [42] based on discrete differential geometry [43]. This approach aims at transforming ideal free-form surfaces into buildable structures by assembling a collection of smaller flat or curved polyhedral elements on a supporting mesh with

Figure 1.   Complex free-form architecture can be made buildable by assembling small flat elements on a torsion-free support structure. This image shows part of the Fondation Louis Vuitton building, in Paris, designed by F. Gehry. The free-form walls are made of flat white UHPFRC panels, assembled on a stainless steel support mesh. The large glass sails are made according to the same principle.

Source: Photography by the author.

torsion-free nodes. Steel-and-glass and wood-and-wood are the most frequently encountered combinations, but ultra-high-performance-fiber-reinforced concrete (UHPFRC) panels are also perfectly adapted to free-form architecture (Figure 1).

8. Toward Rebar-Free Concrete

The Achilles' heel of concrete is its relatively low tensile strength. This is why concrete needs reinforcement. Conventional reinforcement with rebars has many advantages. It is an inexpensive and robust technology. It is easy to build on site, with a minimum of training. It provides tensile strength, ductility and crack growth resistance. Whatever the future cement and concrete technologies, the need for reinforcement will probably not disappear soon.

Yet, conventional reinforcement has serious drawbacks, in particular in terms of work force needs and productivity. While robotic placement, bending and welding could be a solution, either on site or in the prefabrication plant [44], other approaches taking advantage of digital technologies have been considered. For instance, digital methods may help in designing structures with minimal levels of tension or even compression-only structures, with much reduced reinforcement needs [45]. Roman vaults are simple examples of compression-only structures. They would stand without any mortar.

Autonomous assembly of the reinforcement units is another, radically different approach [46]. The concept of autonomous assembly may be viewed as a generalization of the concept of self-organization or self-assembly, which has been widely popularized as a key concept in our understanding of life and many other natural phenomena. It is currently intensively explored in the context of fabrication, architecture or structural engineering under various names like "designer matter", "aleatory architecture" or "aggregate architecture". A common underlying assumption is that, with properly designed building blocks, functionality may emerge from the interaction of a large number of building blocks during the assembly process which, in the simplest case, may be totally random.

9. Introducing Robotics in Construction

The construction sector is still far from the massive move toward robotic methods observed in the manufacturing industry. Yet, robots are fully present in architectural research [47] and an increasing number of laboratories, companies or consortia are experimenting and developing robotic methods suitable for either prefabrication or on-site construction with concrete [48]. Formwork-free 3D printing by multilayer addition of extruded layers [49], often called contour crafting after the seminal work of B. Khoshnevis [50], is the most widely explored method. Concrete is deposited by pumping it in a robotically controlled mobile extrusion nozzle and the structure is progressively built by passing the required number of times. This method is applicable on the construction site.

Particle-bed printing, initiated by E. Dini [51], is an interesting alternative. In this method, a layer of dry mortar powder is deposited on a flat surface and a robotic arm sprays a binder (in this case, pure water) where the powder has to be consolidated (for instance, along the future walls of the building). A new layer of powder is deposited on the first layer and the consolidation process is repeated. In the end, the large excess of unconsolidated powder is removed and recycled. Needless to say, this can only be used in prefabrication in a closed environment.

Fast additive manufacturing by extrusion of fresh concrete or mortar — the most widely used method — is still a challenge. It requires an accurate control of the fresh paste rheological behavior and, most important, of its so-called structural build-up kinetics, that is, the rate of yield stress increase once an extruded layer has been laid down [52]. Structural build-up determines the rate at which layers can be piled on top of each other without buckling, while still allowing for good adhesion between layers.

Rapidity of construction is often put forward as the main advantage of robotic construction, but this is far from obvious. When massive production of simple building or infrastructure elements is concerned, classical prefabrication techniques with formworks remain very competitive. The real benefit from using robotic printing technologies is not necessarily in the way (faster, cheaper, etc.) we build, but more likely in *what* we build. Not many quantitative assessments have been made so far, but a key finding is that robotic construction with concrete produces high environmental benefits compared to conventional construction when complex structures are built [53]. With robotic 3D printing technologies, additional complexity can be achieved without additional environmental costs, so the potential benefit of digital fabrication increases proportionally to the level of complexity of the structure. This is the perfect incentive for implementing the smart design methods summarized in Section 7, provided a real benefit is obtained in terms of environmental impact, productivity and/or human wellbeing.

10. Toward a Cradle-to-Cradle Economy of Concrete

The cradle-to-cradle concept relys on two elements: total recycling on the one hand, and vanishing climate change potential, on the other hand. The present state of the cement and concrete business is undoubtedly far from fulfilling these two conditions. However, significant progress is being made on both aspects.

Full recycling of all the components of concrete, including cement, is still a difficult and expensive process, not only for technical reasons but also for lack of organization in the whole value chain, in particular in building deconstruction and demolition waste management. The present state of concrete recycling is essentially limited to the use of concrete lumps from building demolition wastes as road construction materials or as recycled aggregates in new concrete, but things may change quite rapidly. Accelerated carbonation of the concrete lumps with atmospheric carbon dioxide or with flue gases before re-using them is an interesting additional step in the recycling process. It improves the carbon footprint of the material and, simultaneously, increases the hardness of the lumps. More recently, true recycling of the coarse aggregates (that is, crushing concrete lumps and separating and cleaning each aggregate class from the finer fraction attached to it) has been demonstrated at industrial scale. The recycling of fine aggregates (sand) will follow. Separating and recycling the fillers would be a much more tricky operation, due to their intricate association with the binding phase. A better strategy would be to feed the whole fine fraction (hardened cementitious matrix with the fillers in it) in the kiln as substitute for limestone and clay.

As far as the climate change potential of concrete is concerned, the future is less clear. The major contribution is undoubtedly from the Portland cement or, more precisely, from the clinker manufacturing process, with two distinct major sources of carbon dioxide emissions. The first is directly related to the high temperature ($-1450°C$) of the process. This process is now close to a thermo-chemical optimum, with very efficient heat recovery and use. However, the source of that heat is still the burning fossil fuels (pet coke) or combustible organic wastes, which leads to carbon dioxide emissions. The second source of carbon dioxide emissions during clinker manufacturing is directly related to the very chemistry of the clinker formation process, which implies calcium carbonate decarbonation ($CaCO_3 \rightarrow CaO + CO_2$). As long as Portland clinker remains the basis of our cementitious binders, this will remain a baseline for the carbon emissions.

Among the possible ways to move this baseline, the most obvious is to reduce the amount of clinkers in cement, as discussed in Section 5 [30,34], but this has limits either in terms of availability of the substitution materials or in terms of properties of the blend. Carbon dioxide capture and underground storage (CCS) is another option, but with large investment and operation costs. A more radical approach would be to divest from thermo-chemical process routes and to explore room temperature electrochemical routes using renewable electricity. Recent work suggests that this is in principle possible, but the process design is still at a very theoretical stage [54]. An even more radical solution would be to abandon Portland-type cements and to switch to a different — yet to be discovered — chemistry with a better carbon balance and, hopefully, similar binding properties and convenience of use as Portland cement [55]. In this respect, magnesium oxide-based binders derived from magnesium silicates seem to be particularly interesting, especially if hardening by carbonation with atmospheric or industrial CO_2 is implemented [56]. A carbon-negative life cycle is in principle possible, but the industrial feasibility of this remains to be established. All together, in spite of many — sometimes disruptive — possible improvements, we are still far from a decarbonated cement and concrete sector [57].

11. Conclusion

The coming decades are probably not going to be the last ones of concrete's biography. As far as basic facts and anticipated needs are concerned, the future still looks bright. In most situations, concrete is still an incredibly convenient and easy-to-use material. Its binder, Portland-type cement, is a mixture of oxides of the four most abundant elements in the earth's crust, besides oxygen: silicon, aluminum, iron and calcium. It is no surprise that, in a century of rapid growth of the world population and of its infrastructure and housing needs, it became the quasi-default material for any new construction project. This period is not yet over. A large part of the planet is still eager to catch up with what is considered as a right for modernity.

Yet, the environmental urgency makes the GHG emissions of the cement and concrete sector and its extensive use of virgin natural resources no longer acceptable. Coping with these challenges is not only the matter of incremental improvements — the most obvious of them have been summarized in Sections 4–9 — it will require more radical changes (Section 10), both in the way we use concrete and in the way we manufacture the binder.

The organization of a truly circular economy of concrete is the first ingredient of this new construction world. It is on its way, but needs to speed up. To divest from a more than two centuries old energy-hungry way to manufacture a hydraulic binder is a more challenging enterprise. For too long the cement industry has been reluctant to leave its comfort zone. It should not be blamed for that. By constantly improving its high temperature process and by incorporating wastes from other industries without any loss of properties, it succeeded in providing the world with one of the less expensive and most useful materials on Earth. But times have changed and it would be wise now to seriously explore new territories.

References

1. Scrivener, K. L., John, V. M. and Gartner, E. M. (2016). Eco-efficient cements: Potential, economically viable solutions for a low-CO_2, cement-based materials industry. United Nations Environment Program, www.unep.org.
2. Ramage, M. H., Burridge, H., Busse-Wicher, M., Fereday, G., Reynolds, T., Shah, D. U., Wu, G., Yu, L., Fleming, P., Densley-Tingley, D., Allwood, J., Dupree, P., Linden, P.F. and Scherman, O. (2017). The wood from the trees. The use of timber in construction. *Renewable and Sustainable Energy Reviews*. 68, 333–359.
3. World Steel Association (WSA). Steel statistical yearbooks 1978–2016, www.worldsteel.org/steel-by-topic/statistics/steel-statistical-yearbook-.html.
4. U.S. Geological Survey (USGS). Iron and steel statistics and information, https://minerals.usgs.gov/minerals/pubs/commodity/iron_&_steel/mcs-2018-feste.pdf.
5. Geyer, R., Jambeck, J. R. and Law, K. L. (2017). Production, use, and fate of all plastics ever made. *Science Advances*. 3, e1700782. Supplementary material. http://advances.sciencemag.org/cgi/content/full/3/7/e1700782/DC1.
6. U.S. Geological Survey (USGS). Aluminium statistics and information, https://minerals.usgs.gov/minerals/pubs/commodity/aluminum/mcs-2018-alumi.pdf.
7. IEA. (March 2019). Material efficiency in clean energy transitions. International Energy Agency Report, https://www.iea.org/publications/reports/MaterialEfficiencyinCleanEnergyTransitions/.

8. U.S. Geological Survey (USGS). Cement statistics and information, https://minerals.usgs.gov/minerals/pubs/commodity/cement/.

9. CEMBUREAU. (2016). Activity report, https://cembureau.eu/media/1635/activity-report-2016.pdf.

10. WBCSD. (2016). Cement sustainability initiative, getting the numbers right, project emissions report 2014, https://docs.wbcsd.org/2016/12/GNR.pdf.

11. Picon, A. (2006). Architecture and technology: Two centuries of creative tension. In Cohen, J.-L. and Martin Moeller, Jr., G. (Eds.), *Liquid Stone. New Architecture in Concrete*. Birkhäuser, Basel, pp. 8–19.

12. Cohen, J.-L. and Moeller, G. M. (2006). Introduction. In Cohen, J.-L. and Martin Moeller, Jr., G. (Eds.), *Liquid Stone. New Architecture in Concrete*. Birkhäuser, Basel. pp. 6–7.

13. Cohen, J.-L. (2006). Modern architecture and the saga of concrete. In Cohen, J.-L. and Martin Moeller, Jr., G. (Eds.), *Liquid Stone. New Architecture in Concrete*. Birkhäuser, Basel. pp. 20–33.

14. Torres, A., Brandt, J., Lear, K. and Liu, J. (2017). A looming tragedy of the sand commons. *Science Magazine*. 357(6355), 970–971.

15. Collins, P. (2004). *Concrete. The Vision of a New Architecture*. 2nd ed., Montreal & Kingston: McGill-Queen's University Press.

16. Simonnet, C. (2005). *Le Béton. Histoire d'un Matériau, Editions Parenthèses*. Marseille.

17. Courland, R. (2011). *Concrete Planet: The Strange and Fascinating Story of the World's Most Common Man-Made Material*. Amherst, NY: Prometheus Books.

18. Freyssinet, E. (September 15, 1928). L'amélioration des constructions en béton armé par l'introduction de deformations élastiques systématiques. *Le Génie Civil*. 240, 254–257.

19. Okamura, H. and Ozawa, K. (1995). Mix-design for self-compacting concrete. *Concrete Library of JSCE*. 25, 107–120.

20. Richard, P. and Cheyrezy, M. (1995). Composition of reactive powder concrete. *Cement and Concrete Research*. 25, 1501–1511.

21. Wangler, T., Lloret, E., Reiter, L., Hack, N., Gramazio, F., Kohler, M., Bernhard, M., Dillenburger, B., Buchli, J., Roussel, N. and Flatt, R. (2016). Digital concrete: Opportunities and challenges. *RILEM Technical Letters*. 1, 67–75.

22. Barbosa, F., Woetzel, J., Mischke, J., Riberinho, M. J., Sridhar, M., Parsons, M., Bertram, N., Brown, S. (February 2017). Reinventing construction: A route to higher productivity. McKinsey Report.

23. Van Damme, H. (2018). Concrete material science: Past, present, and future innovations. *Cement and Concrete Research* 112, 5–24.

24. Chateau, X. (2012). Particle packing and the rheology of concrete. In Roussel, N. (Ed.), *Understanding the Rheology of Concrete*. Cambridge: Woodhead Publishing. pp. 117–143.

25. Vernet. (2004). Ultra-durable concretes: Structure at the micro- and nanoscale. *Materials Research Society Bulletin*. 29, 324–327.

26. de Larrard, F. (1999). Concrete mixture proportioning: A scientific approach, modern concrete technology series, N°9. E & FN SPON, London.

27. Young, B. A., Hall, A., Pilon, L., Gupta, P. and Sant, G. (2019). Can the compressive strength of concrete be estimated from knowledge of the mixture proportion? New insights from statistical analysis and machine learning methods. *Cement and Concrete Research*. 115, 379–388.

28. Ziolkowski, P. and Niedostatkiewicz, M. (2019). Machine learning techniques in concrete mix design. *Materials* 12, 1256. Doi: 10.3390/ma12081256.

29. Daminelli, B. L., Kemeid, F. M., Aguiar, P. S. and John, V. M. (2010). Measuring the eco-efficiency of cement use. *Cement and Concrete Composites*. 32, 555–562.

30. Scrivener, K. L., John, V. M. and Gartner, E. M. (2018). Eco-efficient cements: Potential economically viable solutions for a low-CO_2 cement-based materials industry. *Cement and Concrete Research*. 114, 2–26.

31. Lothenbach, B., Scrivener, K. and Hooton, R. D. (2011). Supplementary cementitious materials. *Cement and Concrete Research*. 41, 1244–1256.

32. Matschei, T., Lothenbach, B. and Glasser, F. P. (2007). The role of calcium carbonate in cement hydration. *Cement and Concrete Research*. 37, 551–558.

33. Scivener, K., Martirena, F., Bishnoi, S. and Maity, S. (2018). Calcined clay limestone cements (LC^3). *Cement and Concrete Research*. 114, 49–56.

34. Miller, S. A., John, V. M., Pacca, S. A. and Horvath, A. (2018). Carbon dioxide reduction potential in the global cement industry by 2050. *Cement and Concrete Research*. 114, 115–124.

35. Rudofsky, B. (1987). *Architecture Without Architects – A Short Introduction to Non-Pedigreed Architecture*. Albuquerque: University of New Mexico Press.

36. Van Damme, H. and Houben, H. (2018). Earth concrete. Stabilization revisited. *Cement and Concrete Research*. 114, 90–102.

37. Moevus, M., Jorand, Y., Olagnon, C., Maximilien, S., Anger, R., Fontaine, L. and Arnaud, L. (2016). Earthen construction: An increase of the mechanical

strength by optimizing the dispersion of the binder phase. *Materials and Structures*. 49, 1555–1568.

38. Landrou, G., Brumaud, C., Winnefeld, F., Flatt, R. J. and Habert, G. (2016). Lime as an anti-plasticizer for self-compacting clay concrete. *Materials*. 9, 330. Doi: 10.3390/ma9050330.

39. Pawlin, M. (2011). *Biomimicry in Architecture*. London: RIBA Publishing.

40. Vincent, J. (2010). New materials and natural design. In Allen, R. (Ed.) *Bulletproof Feathers. How Science Uses Nature's Secrets to Design Cutting-Edge Technology*. Chicago and London: University of Chicago Press. pp. 132–171.

41. Zhang, W., Zhu, J. and Gao, T. (2016). *Topology Optimization in Engineering Structure Design*. London: ISTE Press Ltd, Oxford: Elsevier Ltd.

42. Pottmann, H., Eigensatz, M., Vaxman, A. and Wallner, J. (2015). Architectural Geometry. *Computers & Graphics*, 47, 145–164.

43. Bobenko, A. I. and Suris, Y. B. (2008). *Discrete differential geometry: Integrable structure*. Graduate Studies in Mathematics, vol. 98, Providence: American Mathematical Society.

44. Asprone, D., Menna, C., Bos, F. P., Salet, T. A. M., Mata-Falcón, J. and Kaufmann, W. (2018). Rethinking reinforcement for digital fabrication with concrete. *Cement and Concrete Research*. 112, 111–121.

45. Block, P., Van Mele, T. and Rippmann, M. (2017). Compressive assemblies. *Architectural Design*. 248, 104–109.

46. Tibbits, S. (2017). From automated to autonomous assembly. In Tibbits, S. (Ed.), *Autonomous Assembly, Architectural Design*, Vol. 87, pp. 7–15.

47. Gramazio, F. and Kohler, M. (2008). *Digital Materiality in Architecture*. Baden, Switzerland: Lars Müller Publishers.

48. Flatt, R. J. and Guest, T. W. (Eds.). (October 2018). Special issue of Cement and Concrete Research. Vol. 112.

49. Buswell, R. A., Leal de Silva, W. R., Jones, S. Z. and Dirrenberger, J. (2018). 3D printing using concrete extrusion: A roadmap for research. *Cement and Concrete Research*. 112, 37–49.

50. Khoshnevis, B., Hwang, D., Yao, K.-T. and Yeh, Z. (2006). Mega-scale fabrication by contour crafting. *International Journal of Industrial Systems Engineering*. 1, 301e320.

51. Lowke, D., Dini, E., Perrot, A., Weger, D., Gehlen, C. and Dillenburger, D. (2018). Particle-bed 3D printing in concrete construction – Possibilities and challenges. *Cement and Concrete Research*. 112, 50–65.

52. Reiter, L., Wangler, T., Roussel, N. and Flatt, R. J. (2018). The role of early age structural build-up in digital fabrication with concrete. *Cement and Concrete Research*. 112, 86–95.

53. Agusti-Juan, I., Müller, F., Hack, N., Wangler, T. and Haber, G. (2017). Potential benefits of digital fabrication for complex structures: Environmental assessment of a robotically fabricated concrete wall. *Journal of Cleaner Production*. 154, 330–340.

54. Ellis, L. D., Badel, A. F., Chiang, M. L., Park, R.J.-Y. and Chiang, Y.-M. Toward electrochemical synthesis of cement – An electrolyzer-based process for decarbonizing CaCO3 while producing useful gas streams. PNAS.

55. Gartner, E. and Sui, T. (2018). Alternative cement klinkers. *Cement and Concrete Research*. 114, 27–39.

56. Favier, A., Scrivener, K. and Habert, G. (2019). Decarbonizing the cement and concrete sector: Integration of the full value chain to reach net zero emissions in Europe. *IOP Conference Series: Earth and Environmental Science*. 225, 012009.

Part 2

Materials of Promise

Bernadette Bensaude-Vincent

While Part 1 displays materials that have been used and improved over the last millennia, the four materials of this part are typical icons of modernity. These "new materials" became star materials in connection with emerging technologies. In this respect, they exemplify the relational identity of materials. Aluminum changed from precious metal to a mass-product thanks to a process of production (Hall–Heroult) and the availability of cheap electricity. Plastics became ubiquitous as one of the applications of petrochemistry. Over 60 years, silicon has enabled the exponential growth of information technology. Thanks to the semiconductor properties of highly purified silicon, chips of 18 inches of diameter afford a substrate for millions of transistors, thus allowing the miniaturization of electronic devices. And silicon may start a second life as a material for photovoltaic solar panels in energy technology. As for nanocarbons, they are more like young brilliant graduates from high school with star grades. Still they are quickly gaining traction thanks to heavy public and private funding in two emerging sectors: optoelectronics and nanomedicine.

The four biographies in this part also raise ethical considerations about the responsibility of scientists and corporations in the promotion of innovations. All four materials have been advertised as symbols of progress, and promises of a better life. This economy of promises has been favored by the development of marketing and ad campaigns and by the increasing pressure from funding agencies to do research with a potential economic impact in the 20th century. In a context of future-oriented innovation policies, scientists get into the habit of making more firmly quite tenuous claims that their research will lead to spectacular outcomes. Their exhilarating discourses about the potentials of new materials generated by the fervor for technological innovations proved helpful for securing R&D investments and public support in the 20th century and the early 21st century.

However, this part shows that the promise of superlative outcomes of new materials often prevented the anticipation of environmental and societal impacts of their actual production and consumption. Aluminum and plastic commodities have been and are still mass-produced despite the growing evidence of the flipside of these stars of modern materials. Aluminum's brilliant career is tarnished by the polluting extraction of its ore, the high-energy cost of the manufacturing process, and the financial control of the world production by a few companies since the 1980s. Plastics enabled the mass-production of ephemeral commodities, and the intensive consumption of single-use disposable items. Still the ubiquitous presence of ephemeral plastic products in daily life can no longer conceal the long lifetime of the stuff they are made of. The tons of long-lasting residues spread over the land and the oceans emphasize the contrast between the commercial and the environmental history of this material.

The spectacular ascension of silicon as the signature of the information age, supposedly encouraging the dematerialization of technology, raises similar problems due to the energy cost of the process of purification and the accumulation of digital waste due to the exponential increase of the consumption of computers and smartphones.

Nanocarbons are an exception, because the anticipation of their potential impacts on health, environment and society was an integral part of the promises accompanying R&D. But their future will depend upon technical solutions to the difficulties of standardization as much as on social regulations because they raise challenging risk issues due to their surface of interactions with the milieu where they operate.

By focusing on the ambivalent effects of the economy of promises in materials innovation, this part invites considerations about good practices in innovation policies as broader ethical reflections about the future and intergenerational justice.

Chapter 5

Aluminum

Florence Hachez-Leroy

Artois University, 9 Rue du Temple, 62000 Arras, France

1. Introduction

Aluminum is the most abundant metal in the Earth's crust. It occurs naturally in its oxidized state in a variety of sedimentary rocks such as bauxite. Alumina (Al_2O_3) is itself a material with good mechanical, chemical and thermal qualities, used traditionally in the production of ceramics. Its isomorphic form, beta-alumina, became a promising material for manufacturing solid electrolytes in the 1970s. Alums, hydrated double sulfate salts of aluminum ($Al_2 (SO_4)3nH_2O$), have been used since Roman antiquity in dyeing and tanning processes. However, the story of aluminum as a metallic material capable of replacing conventional metals in a wide range of industrial applications only began in the 19th century, since its production is intimately bound up with the history of electric power. This dependency exemplifies the relational identity of certain materials as natural entities that offer performances only in relation to specific conditions and circumstances. Cheap electricity produces the cheap metal that helped shape the modern world during the 20th century. However, the identity and the image of the metal has changed significantly over the past two centuries. It was initially prepared as a semi-precious metal used in jewelry and silversmithing. Then, when its price dropped, it became a common material widely used in everyday life. The properties of the metal — lightweight, conductive, non-toxic and easily machined — make it an ideal material for transportation and packaging. But if aluminum lived up to its promises throughout the 20th century, this successful career also had a darker side.

2. From Scientific Curiosity to Semi-Precious Metal

The discovery of aluminum was the result of a long journey which began in the 18th century with research dedicated to plant growth and soils [1]. The research movement started in 1792 with Lavoisier and was pursued by Humphry Davy in 1804. The scientific agenda can be seen in the "alumina" entry in *L'Encyclopédie méthodique*, in its 1792 edition: "The desires of chemists are still very far from being satisfied as to what they would like to know about alumina. The intimate nature of this earth is still entirely unknown, like that of other earthy materials" [2 p. 193]. Theodore de Saussure, in 1801, devoted a detailed notice to alumina in the *Journal de physique, d'histoire naturelle et des arts* [3], and explained his observations and his methods of production. Three years later, in his 1804 reference book, *Recherches chimiques sur la végétation,* he mentioned his observations of the presence of alumina in the ashes of various plants on which he had conducted experiments [4].

In the same year, while using a voltaic pile to split common compounds such as alum, a white mineral used since Greek and Roman antiquity for dyeing and tanning processes, Humphry Davy identified the presence of two metals in it: potassium and another one that he failed to isolate as a simple substance. He first named it "alumium" then changed this name to "aluminum" in 1807 before adopting "aluminium" in his *Elements of Chemical Philosophy* in 1812. His hesitation was the source of a persistent spelling confusion despite the chemists' concerns to standardize their nomenclature. The term "aluminium" is recommended today by the International Union of Pure and Applied Chemistry because it retains the suffix "-ium" used for other metals such as sodium or potassium. But this official term is still in competition with "aluminum" adopted by Webster's dictionary, in American language.

Whatever its name, this simple metallic substance is difficult to extract from its ores. It was first isolated in an impure form by Hans Christian Ørsted in 1825. In 1827, Friedrich Wöhler obtained a few grains of pure aluminum by reducing its oxide with potassium. The French chemist Henri Sainte-Claire Deville reproduced Wöhler's experiment using liquid sodium instead of potassium aluminum and succeeded, for the first time, in obtaining significant quantities of the metal [5]. Sainte-Claire Deville's research culminated in the winter of 1853–1854 and he made his first communication to the Paris Academy of Sciences in its session of 6 February 1854. In March 1854, for the first time in the world, an aluminum blade was presented to the public. In his cover letter to the Academy, the scientist specified that tests had been carried out on the metal for three months [6]. Sainte-Claire Deville's success was carefully orchestrated and the greatest scientists of the time were recruited to applaud it [1]. In August 1854, he performed his experiment before a number of fellow chemists including Jean-Baptiste Dumas, Louis-Jacques Thenard, Jean-Baptiste Boussingault and Théophile-Jules Pelouze [7]. Others, such as the German researcher Justus von Liebig, received a sample of the metal. This was a particularly strategic gift: in addition to his qualities as a chemist which were recognized worldwide, especially in the realm of foodstuffs, von Liebig was also a collaborator of Wöhler. So aluminum emerged first of all then as a scientific curiosity.

Sainte-Claire Deville's discovery was also publicized in North America. In 1854, *The Mining Journal* reported on a presentation by Henry Wurtz at a meeting of the New Jersey National History Society:

"[Sainte-Claire Deville] has succeeded in obtaining pure aluminum in considerable masses, and ascertaining thoroughly its properties. […] It may safely be predicted, allowing Deville's assertions to be correct, and they have been confirmed by a Commission appointed by the French Academy of Science to investigate the matter, that some of us now present may yet live to eat with forks and spoons composed of aluminum, on aluminum dishes, food cooked in aluminum utensils upon an aluminum stove, and possibly while seated upon aluminum chairs in houses or ships composed of the same metal" [8].

It was still a long way, however, from this scientific discovery to the industrial production of aluminum utensils [9]. In 1854, and with some financial support from Emperor Napoleon III, the Javel factory in Paris produced the first blades of aluminum, which were as expensive as silver

Figure 1. Table center-piece with putti (1858, coll. Compiègne, larg. 40 cm, Copyright © C2RMF).

ones. At the Paris International Exhibition of 1855, a number of objects, such as metal tumblers designed by the goldsmith and tableware company Christofle, and gradually crafted out of the small quantities of metal produced, were presented to the public. This firm also made a splendid epergne ordered by the Emperor for the imperial palace of Compiègne (Figure 1).

In a fiction *From the Earth to the Moon, A Direct Route in 97 Hours 20 Minutes,* French novelist Jules Verne accurately described the tentative design of an enormous space gun to launch three people onto the moon. To the enthusiast engineers in charge of the project, aluminum appeared as the best material for the projectile.

> This valuable metal possesses the whiteness of silver, the indestructibility of gold, the tenacity of iron, the fusibility of copper, the lightness of glass. It is easily wrought, is very widely distributed, forming the base of most of the rocks, is three times lighter than iron, and seems to have been created for the express purpose of furnishing us with the material for our projectile [10, p. 38].

3. Everywhere in Daily Life

Aluminum first came into existence then as a substitute for precious metals used for luxury items. By the end of the 19th century, however, the industrial production of the metal had witnessed its timid beginnings. The Salindres factory, in the Gard department in the South of France, located near bauxite mines, had already manufactured 45 tons of aluminum by 1889. Because of its gold color, the aluminum-bronze alloy was particularly successful [11]. However, aluminum artifacts were still the hallmarks of the salons of the French bourgeoisie. It was to become a common, mass-produced material only in the early years of the 20th century, when the obstacles along the route toward a commercially viable material were finally removed.

In 1886, two young men working independently, Paul Héroult in France and Charles-Martin Hall in the United States, found a way to overcome a major obstacle in the electrolytic process used to reduce aluminum oxide. To avoid the formation of aluminum hydroxide when the oxide was dissolved in water, they used molten cryolite as a solvent and electrolyzed the molten salt. The scaling-up of the electrolysis of molten cryolite still required high temperatures however and considerable power. Therefore, the commercial success of aluminum production was dependent on the availability of a cheap source of energy, hydroelectricity. It was hydroelectricity that made the mass production of aluminum possible. Aluminum plants consequently flourished in specific geographical areas such as the Alps or near the Niagara Falls where water falls or dammed mountain lakes could provide cheap hydroelectricity.

In 1907, aluminum selling prices fell sharply from 3.1 francs per kilogram to 1.5 francs. Several factors explain this price fall. The Hall–Héroult patents had fallen into the public domain and outsiders had appeared, leading to overproduction. The market, still weak, did not have many outlets for aluminum. But the development of the aluminum-copper-manganese-magnesium alloy by the German Alfred Wilm, between 1906 and 1909, was a decisive step in the development of the use of the metal in aeronautics.

During the First World War, in its context of scarcity, aluminum was used as an ersatz for silver and copper and began to look like the metal of the poor. As a cheap metal its uses expanded while its image deteriorated. Fortunately, however, the new metal also catalyzed the creativity of designers [12]. In the interwar period they were generously funded by aluminum companies to improve the image of the metal. Unlike other metals, by means of anodization, aluminum can be colored while keeping the material visible. This characteristic has been widely exploited by designers for decorative purposes and in everyday commodities. The streamline movement in the United States, with Raymond Loewy or Russel Wright, as well as the modernist and avant-garde movements in Europe with Gio Ponti, Robert Mallet-Stevens or Jacques Le Chevallier, managed to change the image of light alloys [13,14]. The products they designed have entered the most famous decorative arts museums in the world.

Transportation has been the most important sector for the development of aluminum alloys since the First World War and continues to be so in 2020 [14]. Aviation, in particular, benefited from the development of Duralumin, a light and resistant aluminum-copper alloy, the sheets of which are quenched to increase their rigidity. For rocketry and aerospace applications, new high-performance alloys have been designed. In the automobile industry, aluminum was introduced as early as 1899 for the *Jamais Contente*, the "never-satisfied", an electric vehicle which could be driven faster than 100 kilometers an hour. Aluminum alloys are still used today for casting engine parts and truck bodies. A few aluminum boats have also been successfully developed, such as ore carriers or sailing boats. Some are famous like *Le Migron* designed by Alfred Nobel in 1892 or Eric Tabarly's *Pen Duik III* in 1967. Yet, steel is still largely dominant in the boat building industry.

Packaging and kitchen utensils represent the second important sector for aluminum applications. During the interwar years, it was developed for pots and camping gear and in the 1950s it became an essential material in the kitchen. Its ability to go from very cold to very hot without alteration opened up the freezing and baking markets. Flexible packaging took advantage of its waterproofness and its ability to be rolled into extremely thin sheets. During the Second World War, multi-layer packaging was developed for the US military and subsequently transferred to the civilian market. Rolled with paper and plastic or coated with a varnish, aluminum is present in fruit juice cartons, toothpaste tubes and drink cans.

Aluminum also found some niche applications in buildings. Decorations of the Bordeaux Cathedral in France, the pyramidal cap of the Washington Monument, the so-called *Eros* statue of 1893 at Piccadilly circus in London or the roofing of the dome of the church of San

Gioacchino in Prati, in Rome, were among the first architectural projects to use aluminum. Later, the excellent mechanical properties of aluminum manufactured by the Kaiser company after the Second World War enabled Buckminster Fuller to design new architectural forms for houses and domes. Its capacity to be cast in curved shapes meant that aluminum could also be used for door frames, windows and façades with innovative shapes. This was the case, for example, of the decorative façade of the *Die Zeit* newspaper office in Vienna, designed by Otto Wagner in Vienna in 1902. In skyscrapers, aluminum has become a key material for façades on account of its lightness and strength. Jean Prouvé, a French metal worker and self-taught architect and designer, pioneered the construction of all-aluminum houses and tropical pavilions using techniques inspired by aircraft construction [14].

4. A Monopolistic Industry

The successful expansion of aluminum in our everyday urban landscape was made possible by the regular increase in its industrial production. In 1918, world production stood at 131,000 tons, in 1968 at about 10 million tons. It reached 20 million tons in 1996, then doubling again and in less than 15 years: 40 million tons in 2010 and 63.7 million tons in 2019. China has become the leading producer with more than half the world's production. In 2018, India and Russia were in second place, each producing 3.7 million tons. The United States is now in tenth position with 890,000 tons. This spectacular expansion from the second half of the 20th century is related to the price of electricity and to the fact that multinational companies producing aluminum can jurisdiction-hop to wherever they find the best conditions.

Figures for the consumption of aluminum show similar increase, with marked accelerations during the periods of war. For primary aluminum, it was estimated at 196,000 tons in 1918, its highest historical level. It exceeded one million tons consumed in 1943 (1.7 million) of which more than half was consumed by the United States. Consumption then reached 10 million tons in 1971 and 20 million in 2000. If we add primary and secondary aluminum — aluminum that is made from recycling — consumption reached 38 million tons in 2000. Primary aluminum is produced from the electrolysis of alumina; secondary aluminum is obtained by recycling recovered waste such as cans, pots, planes, etc. Recycling developed after the Second World War in a context of scarcity.

The mass-production and consumption of aluminum is based on a specific economic model. In addition to cheap electricity, the industrial production of aluminum requires heavy banking investments. The aluminum industry worldwide is a typically capitalistic industry in the hands of few monopolistic companies. From 1886 until the First World War, there were only five: The Aluminum Company of America (Alcoa Inc.) in the United States, Société Electrométallurgique de Froges and Compagnie des produits chimiques d'Alais et de la Camargue (Pechiney, after merger in 1921) in France, Aluminium Industrie Aktien Gesellschaft (AIAG) in Switzerland and the British Aluminium Company Ltd. in the United Kingdom. These companies held the monopoly until the patents fell into the public domain in 1907. After the First World War, the five pioneering companies helped other countries develop their technology. But in most countries, as in Argentina or in Japan, for example, a monopolistic position was the rule rather than the exception [15,16].

In 1901, an international cartel was set up, the Aluminum Association. One of the oldest and strongest economic cartels in the world, it remained more or less effective until the 1950s. The Aluminum Association and the industrial companies fixed the price of the metal and shared out the world market, while keeping their national markets sheltered from competition. In the United States, the Sherman Act, an antitrust law of 1890, forced Alcoa to split from its Canadian subsidiary in 1928. Thanks to a dumping action, the Canadian Alcan company managed to conquer foreign markets in the 1930s.

In the aftermath of the Second World War, and under US influence, the Treaty of Rome in Europe adopted an anti-cartel position that would force the national companies to put an end to this trust economy [17–19]. However, their practices changed only on a superficial level; industrial associations were created to lobby actively with the common market authorities at Brussels. But the introduction of aluminum on the London Metal Exchange in 1978 marked a sharp break in the balance established in the late 1880s by the major Western producers [20]. Henceforth, the sales price of the aluminum ingot was subject to speculation like copper or nickel. This came as a shock, the effects of which were accentuated by the fall of the Berlin Wall and the arrival of low-priced ex-Soviet metal on the Western market. Aluminum prices plunged and the value of aluminum companies fell. For the historical actors, these events opened a period of profound reconfiguration. Most disappeared, acquired by their competitors in more or less friendly takeover operations. In 2020, only the American Alcoa remains from among the five historical companies and world ranking is completely different. After being held for a long time by North Americans and Europeans, leadership is now dominated by Chinese and Russians [21]. Four Chinese companies are among the nine largest producers in the world. The Russian company Russal is in third place, Rio Tinto (Anglo-Australian) in fifth, Emirates global Aluminum (United Arab Emirates) is sixth and Alcoa (USA) eighth followed by the Norwegian Norsk Hydro. Production capacities have exploded, with ever-larger smelters [22]. Only about 40 countries in the world produce primary aluminum.

5. The Dark Side of the Metal

The successful, innovative and surprising applications of aluminum — from jewelry to saucepans and from airplanes to telescope mirrors — all illustrate the metal's success story [12]. Since its early applications, however, the metal has raised concerns about risks and potentially harmful effects. The creation of aluminum production plants sparked protests against their local impact in terms of pollution. Such controversies increased with the intensification of production in the second half of the 20th century, particularly on account of the environmental footprint of the massive exploitation sites of bauxite mines and alumina production [23].

Unlike other metals that are trace elements (copper, iron, zinc), aluminum is not necessary for the human metabolism. Beginning in the middle of the 19th century, its consumption in metallic form (from kitchen utensils) and as a food additive (baking powder) raised the question of the safety of its use in foodstuffs [1]. At the end of the 19th century, this use of aluminum in food production raised two scientific controversies. One concerned the analysis of the material used for the equipment of French and German soldiers, their flasks in particular. Three scientists, two

German and one French, challenged the safety of aluminum on the grounds that the metal was corroded by the liquids in the flasks. These arguments were refuted by the rest of the scientific community, not unreasonably considering the nature of the material studied and the conditions of the analysis: samples came from objects made of metal obtained from the early days of industrial production when poorly controlled technical parameters would leave impurities.

A second controversy in the United States opposed the champions of baking powder containing aluminum salts to a group claiming that this baking powder was toxic. The latter included producers of aluminum-free baking powder as well as independent researchers convinced of its toxicity. It was the longest food controversy in American history. Theodore Roosevelt set up a special committee, the Remsen board, responsible for assessing the potential risks of aluminum. For the first time brief clinical trials were carried out on humans. Other trials, on dogs and rats, led to the deterioration of the animals' healths and, in some cases, to their death. These observations made it possible to record the neurotoxic effects of aluminum when ingested in large doses, as well as the presence of the metal in various organs such as the liver, the spleen and the brain. Nevertheless, these results did not prevent the use of aluminum in baking powder.

During the interwar period, a third controversy took place in Great Britain. A general practitioner, Robert Montague Le Hunte Cooper, observed major signs of chronic fatigue among some of his patients, along with joint pains and dizziness. He also noted that his patients' condition would improve when they stopped using aluminum kitchen utensils. He delved into the scientific literature on the issue and went on to perform laboratory experiments. But he did not manage to assert his opinion successfully in face of the opposition marshalled by the industrialists defending the metal's innocuousness. In both the American and British controversies, the aluminum industry reacted strongly, even sponsoring publications by renowned scientists refuting the conclusions unfavorable to aluminum. This practice, typical of industrial lobbies as "merchants of doubt", undermined public trust in science but did not affect the production and consumption of aluminum in the food industries [24].

The year 1986 marked a turning point in the history of concerns over aluminum after two British researchers identified "senile plaques" comprising an identical distribution of aluminum and silicon in the brains of patients affected by Alzheimer's disease. Today the role of aluminum in encephalopathies is discussed in scientific publications worldwide. Many mainstream newspapers and daily and weekly publications took up the conclusions of this work, emphasizing its sensationalism and declaring more or less peremptorily that aluminum was the cause of the epidemic of Alzheimer's disease.

By the end of the 20th century, it was established that aluminum plays an important role in several diseases: breast cancer, Crohn's disease, autism and "macrophagic myofasciitis". For the latter, investigations carried out on this uncommon inflammatory disorder of muscle showed the presence, in the macrophages of tissue lesions, of specular inclusions composed of aluminum hydroxide. The phenomenon is related to the use of aluminum as a vaccine adjuvant, in order to improve the immune response of inactivated vaccines. The hypothesis proposed is that chronic fatigue syndrome could be caused by the presence of pathogenic agents or toxic compounds that have immunostimulant effects, provoking an incessant stimulation of the immune system. According to this research, chronic fatigue syndrome is linked to aluminum salts.

The widespread use of aluminum in food and agriculture, cosmetics, pharmaceutical industries and water purification, as well as of course in transportation and daily household objects, has multiplied the sources of exposure to the metal, to the point that aluminum particles are now present in the air. Of course, aluminum, as mentioned, is the third constituent element of the Earth's crust (8.1%) after oxygen and silicon, and before iron (5%), but the rate of exposure of the human body to its possible contamination is today way above the levels seen in the 19th century.

Air pollution (fluorine) from aluminum factories aroused protests from the early 20th century, but conflicts were generally settled by financial compensation for neighboring farmers [25]. The considerable increase in production after the Second World War caused enormous damage to the surrounding flora and fauna and gave rise to occupational diseases. Pressure from environmental movements and from public opinion generally has led manufacturers to find solutions to reduce gas emissions, such as covering the electrolytic cell with recycled gases and the automation of their alumina supply. The situation has improved significantly, but there remains the problem of the huge amounts of electricity consumed to produce aluminum. Although hydroelectricity has been replaced by thermal or nuclear electricity with the subsequent location of factories near deep-water harbors, in the 21st century, the establishment of gigantic factories near oilfields raises serious questions about their carbon footprint.

Finally, upstream from metal production itself, the problems of bauxite mines and alumina factories are just as intractable. With prehistoric caves destroyed by Rio Tinto in Australia or threats to sacred sites in India, the demand for bauxite generates strong environmental conflicts and significant impact on the landscape [26,27]. Likewise, the inability of industrialists to find a solution for recycling the red sludge from alumina production using the Bayer process[1] leads to intolerable situations from an environmental point of view. The on-going dumping of this toxic sludge in the Mediterranean, off the coast of the Gardanne factory in the South of France, for more than 50 years, and the open-air storage of the sludge with possible industrial accidents as in Ajka (Hungary, 2010) are further illustrations of the environmental risks associated with this industry.

6. Conclusion

Although aluminum oxide has been used since ancient times, the affordances of the metal remained unknown until the mid-19th century. It is now ubiquitous in our daily lives and almost indispensable for a number of applications. The Hall–Héroult process together with cheaper electricity enabled the mass production and consumption of aluminum. Until the 1950s, industrial companies developed and consolidated their niches on the market. In the second half of the 20th century, the cartel system

---

[1]The Bayer process involves three steps: first, crushed bauxite is mixed in a solution of sodium hydroxide (caustic soda) at a temperature of 150–200°C; second, after cooling, this mixture is seeded with crystals to precipitate aluminum hydroxide, leaving an insoluble residue, called red mud or red sludge; third, after washing, the hydroxide is heated in a kiln in order to drive off the water and produce several grades of granular or powdery alumina, including activated alumina, smelter-grade alumina and calcined alumina.

gradually eroded the quotation of aluminum on the London metal exchange market. From 1978, finance took control of the industrial companies.

In addition, the promises of aluminum alloys are seriously compromised by environmental issues. Adverse effects are visible all along the value chain, from the extraction sites to waste management. Presented as the metal of modernity and with a performance capable of being metamorphosed by color and surface treatments, throughout its history it has been an ambivalent material and it remains so today.

References

1. Hachez-Leroy, F. (2019). *Menaces sur l'alimentation, Emballages, colorants et autres contaminants alimentaires, XIXe–XXIe siècles*. Tours: PUFR/PUR.
2. Fourcroy, A.-F., Maret, H. and Guillot-Duhamel, J.-P. (Eds.) (1792). *Encyclopédie méthodique, Chimie, pharmacie et métallurgie. T. 2*. Paris: Panckoucke Libraire.
3. Saussure, T. de. (1801). *Journal de physique, d'histoire naturelle et des arts*. (II). tome LII, Paris, J.-J. Fuchs Libraire.
4. Saussure, T. de. (1804). *Recherches chimiques Sur la végétation*. Paris: Vve Nyon Libraire.
5. Paquot, C. (2005). *Henri Sainte-Claire Deville. Chimie, recherche et industrie*. Paris: Vuibert.
6. Sainte-Claire Deville, H. (January 1855). Recherches sur les métaux et en particulier sur l'aluminium et sur une nouvelle forme de silicium. Mémoire présenté à l'Académie des sciences le 14 août 1854. Note additionnelle pour servir de complément au précédent mémoire. *Annales de chimie et de physique*. 43(3), 5–36.
7. Plateau, J. and Nichols, S. (2001). Le livre d'or de l'usine de la Glacière, 1856. *Cahiers d'histoire de l'aluminium*. (28), 7–22.
8. [Anonymous]. The mining magazine devoted to mines. (1854, July–December). *Mining Operations, Metallurgy, &c*. (3), 234.
9. Renaux, T. (2017). L'aluminium au XIXe siècle. Une industrie aux pieds d'argile, entre chimie et métallurgie (1854–1890). [dissertation], Paris: EHESS.
10. Verne, J. (1874). *From the Earth to the Moon*. New York: Scribner.
11. Charpy, M. (2010). Le théâtre des objets. Espaces privés, culture matérielle et identité bourgeoise. Paris, 1830–1914. [PhD. dissertation], Tours:

Université François-Rabelais. Charpy, M. (2017). Du salon au cimetière. Les usages sociaux de l'aluminium au XIXe siècle. In Fridenson, P. and Hachez-Leroy, F. (Eds.) *Aluminium, matière à création, XIXe–XXIe siècles*. Tours: PUFR. pp. 49–60.

12. Nichols, S. (Ed.) (2000). *Aluminum by Design*. Pittsburgh: The Carnegie Museum of Art.

13. Schatzberg, E. (June 2003). Symbolic culture and technological change: The cultural history of aluminum as an industrial material. *Enterprise & Society*. (4/2), pp. 226–271.

14. Fridenson, P. and Hachez-Leroy, F. (Eds.). (2017). *Aluminium, matière à création, XIXe–XXIe siècles*. Tours: PUFR.

15. Rougier, M. (2017). The challenges of the aluminium industry in Argentina (1930–2010). In Fridenson, P. and Hachez-Leroy, F. (Eds.) *Aluminium, matière à création, XIXe–XXIe siècles*. Tours: PUFR. pp. 95–104.

16. Miwa, H. (2017). The rise and decline of smelters and the success of aluminium fabricators in Japan. In Fridenson, P. and Hachez-Leroy, F. (Eds.) *Aluminium, matière à création, XIXe–XXIe siècles*. Tours: PUFR. pp. 105–114.

17. Hachez-Leroy, F. (1997). Stratégie et cartels internationaux en Europe, 1901/1981. In Grinberg, I. and Hachez-Leroy, F. (Eds.) *Industrialisation et sociétés en Europe occidentale de la fin du XIXe siècle à nos jours, L'âge de l'aluminium*. Paris: Armand Colin. pp. 164–174.

18. Hachez-Leroy, F. (Ed.). (2003). The European Aluminium Industry (1945–1975). *Cahiers d'histoire de l'aluminium*. Hors-série No. 1.

19. Bertilorenzi, M. (2018). *The International Aluminium Cartel*. New York: Routledge.

20. Mouak, P. (2010/I–2). L' aluminium au London metal exchange: les spécificités institutionnelles et financières d'un marché à terme. *Cahiers d'histoire de l'aluminium*. (44–45), 106–123. Doi: 10.3917/cha.044.0106.

21. World Aluminium [Internet]. (February 2013). Carmine N. The global aluminium industry, 40 years from 1972. London: International Primary Aluminium Institute. p. 27. http://www.world-aluminium.org/media/filer_public/2013/02/25/an_outlook_of_the_global_aluminium_industry_1972_-_present_day.pdf.

22. Gagné, R. and Nappi, C. (July/August 2000). The cost and technological structure of aluminium smelters worldwide. *Applied Econometrics*. 15(4), 417–432.

23. Sheller, M. (2014). *Aluminum Dreams. The Making of Light Modernity*. Cambridge (Mass.): The MIT Press.

24. Oreskes, N. and Conway, E. (2010). *Merchants of Doubt: How a Handful of Scientists Obscured the Truth on Issues from Tobacco Smoke to Global Warming*. New York: Bloomsbury Press.
25. Hachez-Leroy, F. (2013). Aluminium in health and food: A gradual global approach. *European Review of History*. 20(2), 217–236.
26. Padel, F. and Das, S. (2010). *Out of This Earth: East India Adivasis and the Aluminium Cartel*. Delhi: Orient BlackSwan.
27. Padel, F., Serper, Z. and Salaman, N. (2017). Bauxite battles, aluminium reflections and the forced industrialisation of indigenous India. In Fridenson, P. and Hachez-Leroy, F. (Eds.) *Aluminium, matière à création, XIXe–XXIe siècles*. Tours: PUFR. pp. 329–346.

Chapter 6

Plastics

Bernadette Bensaude-Vincent

Université Paris 1 Panthéon-Sorbonne, France

1. Introduction

Plastics are mundane and familiar materials. While we may not be able to individually name all the members of the big family of synthetic polymers known as plastics — vinyl, polypropylene (PP), polyethylene (PE), polystyrene (PS), acrylonitrile butadiene styrene (ABS), polyethylene terephthalate (PET), polyester (PES), polyamides (PA), polyvinyl chloride) (PVC), polyurethane (PU)… — we all know plastics in general. They are ubiquitous, deeply infiltrated in all aspects of our daily lives from our infancy (babies' soothers and toys) to our adult life (cups, pens, telephones, computers) and they are also part of the major infrastructures enabling transportation (inside of cars and airplanes) and communication (coating of underground or submarine cables). We were born in the plastic age.

While the plastic age is the hallmark of the 20th century, we are still in it in the 21st century. Over the past 70 years, the global annual production ramped up from 1.5 million tons in 1950 to 35 million tons in 1970 and 381 million tons in 2020 [1]. The plastics production follows a continuous growth with a compound annual rate of about 8.6% from 1950 to 2015. Plastics are manufactured everywhere around the world, but in 2019 about one-fourth of the global production came from China. These figures are just one indication that the age of plastics is not coming to an end any time soon.

Yet, a number of insistent matters of concern darken the success story of plastics. Plastics have a lot of useful applications, but their production chain causes many hazards and potential damages. As intimate companions to our lives, plastic commodities have a shorter life than pets. The life story of plastic bags and bottles entangled with human daily lives is only one episode in the lifespan of the stuff they are made of. This discrepancy leads to the accumulation of plastic waste. Despite years of campaigns for a ban on single-use plastic items, the global health crisis amplifies their consumption, thus putting extra-pressure on the circuits of waste management.

The relationship between the properties of plastic as a chemical material and the social-economical history of plastic products is the focus of this chapter. It stresses the contrast between the image of plastics as ephemeral commodities and the lifetime of the materials locked up in them, by looking into the black box of their production and supply chains.

2. Materials of a Thousand Uses

Plastics are produced by chemical synthesis. Indeed not all plastics are synthetic polymers — celluloid, for instance, is made from camphor and nitrocellulose — and all synthetic polymers are not plastic [2]. However, plastics are viewed as artificial products because they came into existence as cheap substitutes for expensive natural materials. Celluloid was initially manufactured by John and Isaiah Hyatt in the 1870s as a substitute for ivory for billiard balls. It was described as a "chameleon material" that could be used for making various things, such as combs, buttons, collars and cuffs. It was viewed as a cheap, nasty, deceptive imitation of natural materials until it was used for making roll films. But while celluloid generated the new photography technology, less flammable substitutes were actively sought out [3].

Leo Baekeland, who filed a patent for the first polymer made from synthetic components in 1907, quickly recognized that he should not market the material made through the condensation reaction of phenol with formaldehyde as an imitative substitute. When the Bakelite Corporation was formed in 1922, bakelite was marketed as a "material of a thousand uses", with a "Protean adaptability" for electric appliances, radio sets, automobiles [4]. A decade later, the campaign orchestrated in the 1930s to promote nylon, the new polyamide synthetic fiber 6–6 invented by William Carothers in DuPont's laboratories, was an attempt to break with the image of synthetics as cheap substitutes for natural materials. The term nylon was selected after months of debate because it avoided all connotations of an artificial substitute for silk [5].

The promotion of plasticity and malleability as the major feature of synthetic polymers is based on their process of production. Plastics are molded. The process of polymerization initiated by bringing the raw materials together and heating them is not separate from molding and shaping. Wood and metals pre-exist the action of shaping them: wood is carved, sculpted or laminated; metals are ductile and malleable when molten at high temperature and can be cast or stamped in a press to form components into the desired size and shape. By contrast, plastics are synthesized and shaped simultaneously. In more philosophical terms, matter and form are generated in one single gesture. This specific process relies on the ability of carbon atoms to form covalent bonds with other carbon atoms or with different atoms. Thus, a chain of hundred carbon atoms can make a single macromolecule. The resulting thermosetting polymers are rigid, with remarkable mechanical properties. They are lightweight, have a high-strength-to-weight-ratio, and high thermal and electrical-insulation properties; they are corrosion resistant, and remain bio-inert. Soon a second generation of polymers came on to the market. As they form weaker chemical bonds, they can be reheated, melted and reshaped. Polyethylene, polypropylene, polyester and PVC — are kinds of super-plastics less rigid than earlier thermosetting polymers.

They had already extended the spectrum of applications of synthetic polymers, when the addition of reinforcing fibers in the polymer matrix helped plastics replace more traditional materials such as glass, wood, steel and even concrete. In the mid-20th century, glass fibers were added to reinforce plastics for military applications such as boats and aircrafts, or civil applications such as electric insulators and tankers [6]. In the 1960s, in the Cold War context of space and military programs, the introduction of long and higher-modulus carbon fibers generated a new class of composite materials designed for a specific task in a specific environment. While reinforced plastics were aimed basically at adding the properties of glass fiber to the plasticity of the polymer matrix, composites did reveal new possibilities of designing new materials with never-seen-before combinations of properties like light weight, and high-temperature resistance, for rockets used in severe conditions. These high-tech "materials by design" extended the market of synthetic polymers far beyond the sector of conventional plastics, which are mass-produced with standard specifications.

However, it should be noted that even conventional plastics are to a certain extent objects of design, since their indefinite adaptability depends not only on the production process but also on the use of a number of additives such as plasticizers, fillers, UV stabilizers, anti-oxidants, and the like. Plastics include more than the seven polymer chains identified by the uniform coding system designed by The Society of Plastics Industry for recycling purposes. They are a mixture or blend of various ingredients. The formulation of plastics — the art of mixing various ingredients in the molten state with precise doses — proved crucial to prevent discoloration, bleaching, cracking, for outdoor uses.

3. The Substance of Dreams

In addition to their inner material properties, plastics owe their tremendous success in the 20th century to marketing strategies. The plasticity or "Protean adaptability" that plagued the future of celluloid was promoted as a major advantage afforded by chemistry. William J. Hale announced the "Silico-Plastic Age" as an echo to marketing slogans such as "one plastic a day keeps depression away" [7]. Plastics have been promoted with a lot of corporate advertising and media coverage [8]. In the 1930s, they were celebrated as a driving force toward the democratization of material goods that would provide jobs and feed the market economy thanks to the rapid obsolescence of the mass products. Chemical substitutes were said to spare natural resources. Williams Haynes claimed that: "the use of chemical substitutes releases land or some natural raw material for other more appropriate or necessary employment" [9]. Thus, plastics would contribute to the conservation and protection of nature. Plastic products would also help the emancipation of women, making their lives easier and more enjoyable by simplifying housework with Tupperware and Formica Top Kitchen Tables.

With their shiny, bright-colored and clean appearance, plastic commodities convey the image of the modern world, where most humans get away from the hard, dirty constraints of materiality. American novelist Tomas Pynchon celebrated the values attached to synthetic polymers in his best seller *Gravity's Rainbow*. They embody the modern ideal of humans' emancipation from nature.

> Plasticity has its grand tradition and main stream, which happens to flow by way of DuPont and their famous employee Carothers, known as the Great Synthetist. His classic study of large molecules spanned the decade of the 20s and led directly to nylon, which [...] was an announcement of Plasticity's central canon: that chemists were no longer to be at the mercy of Nature. They could decide now what properties they wanted a molecule to have, and then go ahead and build it. [10, p. 249f]

The tremendous impact of plastics on 20th century culture justifies the phrase "age of plastics". In the 1950s, French philosopher Roland Barthes wrote a short essay "The stuff of alchemy" about plastics in his review of the mythologies of modernity. He praised plastics for their versatility. "Plastics", he wrote, "are like a wonderful molecule indefinitely changing." [11, p. 97] Because plastics can be given virtually any texture, shape and color, they are a pure potential for change and movement. They epitomize the rapid changes occurring in industrial societies after World War II. They connote the magic of indefinite metamorphoses to such a degree that they lose their substance, their materiality, to become virtual reality. Plastics have thus encouraged the dream of an economy of abundance that consumes less and less matter by using cheap and light plastics. Although Barthes witnessed only the debut of the flood of cheap fashionable and disposable products, especially designed to become obsolete after a few uses, he clearly noted the utopia embedded in plastics.

The American cultural historian Jeffrey Meikle argues that the use of plastics in everyday objects, from Bic pens to razors, telephones and credit cards, gradually fashioned a dream world, close to Disney world. The daily experience of plastics transformed American culture:

"Increasingly that culture was seen as one of plasticity, of mobility, of change, and of open possibility for people of every economic class" [12, p. 45].

As plastics changed the people's material environment, they also reshaped people's mentalities and tastes. In the 1960s, plastics penetrated the world of arts. Painters, sculptors, architects, home and fashion designers used a variety of plastic materials or fiber-glass composites in artworks. "Plastic epitomized the financial freedom ad future orientation of the era. [...] The rapid space of change in pop culture was accompanied by a whole spectrum of new products and surface textures in everything from latex clothing to disposable inflatable furniture" [13, p. 25]. Andy Warhol, the iconic figure of New York pop art, famously asserted "I love plastic". The counter-culture movement, which criticized the American way of life, used the term 'plastic' as a derogatory term meaning fake. "Plastic people" referred to superficial and inauthentic persons whose lives were driven by a passion for consumption and change [12].

4. "The Substrate of Advanced Capitalism"

Jeffrey Meikle noted the first plastics were more "commercial than scientific" innovations [12, p. 5]. Indeed, later on DuPont and other chemical companies conducted scientific research on synthetic polymers that led to important innovations such as nylon in the 1930s and the Kevlar, a reinforcing fiber of exceptional strength and stiffness designed by Stephanie Kwolek, in the 1980s. Despite decades of intensive R&D, it remains that the proliferation of plastics was above all driven by a quest for profit and economic growth. Heather Davis argues that plastics are "the substrate of advanced capitalism" because they proved to be key agents to generate economic value and market development [15]. They have served economic growth by encouraging mass-consumption because they are cheap and affordable. They have favored global trade not only because they are light-weight and easily transportable but also because plastic containers and plastic bags accompanied the emergence of new networks of distribution of commodities in retail and wholesale which require plastic wrap for improving the stackability of goods [16]. In addition, plastic allowed the boom of payment cards [17]. It considerably expanded the scope of the consumer credit market, which is another way of fostering mass-consumption.

Above all plastics generate economic value because they are "made to be wasted" [18]. The molecular arrangement of polyethylene terephthalate in PET bottles has been designed for specific applications in beverages industry with a view that the bottles would be thrown away in the trash after one use. This case of obsolescence built into the design of the product can be viewed as a remote descendent of the strategy of planned obsolescence recommended by Bernard London in the 1930s to end the great depression [19]. However, the planned obsolescence of plastic bottles differs from the strategy implemented in the US automotive car industry by Alfred Sloan who yearly changed the design of the General Motors cars. The obsolescence of plastic bottles does not inspire a continuous desire of something newer, better or more fashionable. Single-use plastic bottles serve economic growth by securing a continuous demand of standard, uniform products that command the repetition of the same gesture: discarding the item after use. Such items generated a culture of disposability that increases our dependency on fossil fuel.

5. Plastics in the Ocean

By the end of the 1990s, on his way back from a journey between Hawaii and California, Charles Moore, a sea captain turned scientist, reported that he had spotted huge amounts of fragments of plastic floating the clockwise currents of the North Pacific Gyre. He came in touch with Curtis Ebbesmeyer, an oceanographer who investigated the paths of ocean currents materialized by the trajectories of objects tossed into the water. In 1992, he studied the currents in the North Pacific by tracking 29,000 plastic bathtub toys (ducks, turtles…) released from the containers of a ship bound from Hong Kong to Takoma during a storm. As plastics persist at the sea surface and can be transported over long distances by converging currents before accumulating in the patch, it became clear that they are resilient in the marine environment. The estimation was that 100 million tons of plastic litter accumulated in what came to be known as the Great Pacific Ocean Garbage Patch.

The image of a continent of plastics soon became popular and successfully alerted the public about the pollution caused by the massive production and consumption of plastics. However, the metaphor of the continent is inadequate and far too optimistic because the plastic items washed into streams and rivers on their way to the oceans do not form a visible and well-delineated solid bulk that could be cleaned up or used as a platform. Plastic wastes make up a soup of tiny plastic debris suspended in the surface water. While a continent of plastics could be manageable, a soup of invisible debris distributed all over the seas and oceans cannot easily be cleaned up and raises a global environmental problem.

Following scientific expeditions — such as the Spanish 2010 Malaspina circumnavigation expedition or the eleven expeditions conducted by the Tara Foundation — worrisome data circulate on the web. The Great Pacific Garbage Patch — that is the high-concentration zone of the North Pacific Ocean — is said to cover approximately 1.6 million square kilometers. Plastic has been found even in the Arctic Ocean, which contrary to the North Pacific Ocean is far from highly populated coastal zones. 88% of the sea surface is said to hold suspended plastic debris. The plastic soup results from our consumption of single-use plastic items. One-third of the plastic production is discarded within a year after being manufactured as disposable items of packaging [20], which means that 8–14 million tons of the plastic products annually manufactured end up in the ocean [1].

On closer observation, the soup is mainly composed of 1–5 mm nurdles, most of them residues of breakdown plastic bags, bottles or toys that have been thrown away. It also includes resin pellets, the plastic particles used as raw materials in the factories of plastic products that have been unintentionally dropped during shipment or treatment and then spread into the sea.

What can be the fate of these metric tons of plastic debris? Although our knowledge about their final destination remains scarce, there is evidence that debris between 0.5 and 5 mm are ingested by marine organisms from small invertebrates, and crustaceans, to whales. Plastic makes up 9% to 35% of the contents of fish stomachs. Jennifer Gabrys describes these marine organisms as "carbon workers", thus insisting on the political dimension of this process of biocencentration of the residues diluted in large volumes of water along the food chain [21]. However the biodegradation of carbon compounds affects the health of these "workers" because of the various additives mixed with the carbon chains [22]. In particular, the resin pellets carry persistent organic pollutants such as polychlorinated biphenyls (PCBs) that can cause severe health damages in wild life such as malformations and cancers when they reach a critical concentration. Moreover as PCBs are endocrine disruptors, they affect the reproductive system of all biota.

While the plastic production continues to grow every year, the volume of plastic debris floating in the open ocean does not increase accordingly [23]. It would be nice to relate this observation to the better management of global plastic waste due to the increasing public awareness of the problem. However, human initiatives should not blind us to other phenomena. It seems that plastic particles below 1mm are deposited in sediments.

6. Technofossils

While public opinion around the world gradually became aware of the environmental pollution caused by plastics, a number of scientists ventured the concept of Anthropocene as the era when humans act as a geological force in 2000 [24]. A Working Group of stratigraphists struggles to mark the end of the Holocene through signatures on the Earth. The indicator should meet three criteria: (i) be artificial, (ii) with a global planetary impact, (iii) and a life-time in the order of geological scale. The Working Group selected two candidates: plastics alongside plutonium as potential markers of the beginning of the Anthropocene. Because of the rise of plastics since the mid-20th century, they are widely distributed in both terrestrial and marine milieus. They thus figure as a key geological indicator of the Anthropocene.

> They are abundant and widespread as macroscopic fragments and virtually ubiquitous as microscopic particles. [...] Plastics are already widely dispersed in sedimentary deposits and their amount is likely to grow several-fold over the next few decades. They will continue to be input into the sedimentary cycle over coming millennia as temporary stores — landfill sites — are eroded [25, p. 4].

Thinking of plastics as technofossils casts new perspective on our use of plastic forks and cellphones. Their residues will be the legacy of the plastic world that we have designed for enjoying a "modern life".

It becomes clear that the tremendous success of plastics is at the cost of two symmetrical blindnesses about their impact, upstream and downstream.

Upstream most consumers of plastic toys and packaging look at them as ex nihilo creatures designed for our consumption and convenience. But they do not come into being out of nothing and nowhere. Around 4 percent of the world oil production serves as feedstock to synthesize plastics and an equal amount is required as energy in the process. Yet, one-third of this production used as packaging is dumped after one use. Our ephemeral toys or gadgets bound to be discarded after a few months are made of residues of ancient lives — algae and plankton — that have been stored and compressed in the Earth's crust for more than 10 million years. In other words, our consumption is like eating time. It is an immoderate chronophagy. In addition, their manufacture requires infrastructure for oil extraction and transportation, that are a major source of geopolitical conflicts. Our path-dependence to oil bears connections with the wars devastating the Middle East and the violence of terrorism.

Downstream, the residues are easy to neglect as long as we think of plastics as immaterial objects. For decades we have been encouraged to believe that plastic commodities cease to exist as soon as they are out of sight. Since they were designed and made for us, once discarded they were supposed to be nothing. We buy all these attractive objects without knowing the labor relations that go into their fabrication and their disposal. Plastics are exemplars of the phenomenon of culturally induced or socially constructed ignorance [26]. Indeed over the past decades huge efforts have been made to recycle plastics and reduce plastic packaging. However, they still have a low impact: today only 30% of post-consumer waste are recycled, about 40% are incinerated for energy recovery and the rest goes to landfills or in the oceans. Plastic recycling is not an easy task [27]. A major obstacle is that common plastics tend to phase separate when melted so that one needs to add virgin materials to reprocess them. The logistics of waste management is complex as it requires separate collection, differentiated flows and processes. While the campaigns for sorting out domestic garbage soothe our consciences, they also contribute to a rebound effect in the global consumption of plastic packaging. The main hurdle is that plastics have a low residual value — dependent on the fluctuations of oil price — so that recycling is by no means a lucrative business like metals recycling. It has generated a waste export market toward emerging countries, with China as the major collector until 2018. But a more sustainable economy of plastics will emerge only when all stakeholders — petrochemical companies, product manufacturers, retailers, consumers, waste managers, NGOs, local and national governments, and regulators — align to reinvent plastics. The step toward "green plastics" is highly desirable. The production of bioplastics is feasible as it relies on the long tradition of carbohydrate and fatty oil chemistry through the use of fermentation, enzymes or genetically modified bacteria. Yet, it is most costly and raises other issues about agricultural production.

7. Conclusion

It is now clear that the tremendous success of plastics rests on our blindness about plastics as materials with a life of their own, independent from the lifespan of the commodities. The dark side of the success story of plastics emphasizes that materials matter. Plastics have been praised for their plasticity: they conveyed a sense of freedom through the control over the material environment, but they ultimately proved to be recalcitrant. Over decades the intensive use of plastic commodities has shaped the world we live in and us — our ways of thinking and our worldviews. It has favored a culture of disposable focused on the present and blind to the past and the future of the stuff plastic commodities are made of. However, our present is conditioned by the accumulated traces of past lives, and the future of the Earth will bear the marks of our present. While the manufacturing of plastics destroys the archives of life on the Earth, its waste constitutes the legacy of the 20th century.

References

1. Data from *The Society of Plastic Industry,* https://committee.iso.org/files/live/sites/tc61/files/The%20Plastic%20Industry%20Berlin%20Aug%202016%20-%20Copy.pdf and Plastic in the Ocean Statistic 2020, https://www.condorferries.co.uk/plastic-in-the-ocean-statistics.

2. Rasmussen, S. C. (2018). Revisitng the early history of synthetic polymers: Critiques and new insigths. *Ambix*. 65(4), 356–372.

3. Friedel, R. (1983). *Pioneer Plastic: The Making and Selling of Celluloid*. Madison: University of Wisconsin Press.

4. Brown, A. (1925). Bakelite — What it is. *Plastics*. 1(17), 28–29.

5. Handley, S. (1999). *Nylon: The Story of a Fashion Revolution*. Baltimore: The Johns Hopkins University Press.

6. Mossman, S. and Morris, P. (Eds.) (1994). *The Development of Plastics*. London: The Science Museum.

7. Hales, W. J. (1932). *Chemistry Triumphant: The Rise and Reign of Chemistry in a Chemical World*. Baltimore: Williams & Wilkins & The Century of Progress Exhibition.

8. Rhees, D. J. (1993). Corporate advertising, public relations and popular exhibits: The Case of Du Pont. In Schroeder-Gudehus, B. (Ed.) *Industrial Society and its Museums 1890–1990*. London: Harwood Academic Publishers, pp. 67–76.

9. Haynes, W. (1936). *Men, Money and Molecules*. New York: Doubleday, Doran & Co.

10. Pynchon, T. (1995). *Gravity's Rainbow*. New York: Pynchon Thomas New Edition.

11. Barthes, R. (1972). *Mythologies*. Trans. by Annette Lavers, New York: The Noon-Day Press.

12. Meikle, J. L. (1986). Plastic, material of a thousand uses. In Corn, J. L. (Ed.) *Imagining Tomorrow. History, Technology, and the American Future*. Cambridge: The MIT Press. pp. 77–96.

13. Flor, T. (2017). *Utopias of Plastic*. The Norwegian Art YearBook. Vol. 25, pp. 10–19.

14. Meikle, J. L. (1995). *American Plastic: A Cultural History*. New Brunswick, NJ: Rutgers University Press.

15. Davis, H. (2017). When the soul of the earth is made of plastic. In *Utopias of Plastic*. The Norwegian Art YearBook. Vol. 25, pp. 10–19.

16. Westermann, A. (2013). The material politics of vinyl: How the State, industry and citizens created and transformed West Germany's consumer democrac. In Gabrys, J., Hawkins, G. and Michael, M. (Eds.) *Accumulation: The Material Politics of Plastic*. London: Routledge. pp. 68–86.

17. Deville, J. (2013). Paying with plastic: The credit card. In Gabrys, J., Hawkins, G. and Michael, M. (Eds.) *Accumulation: The Material Politics of Plastic*. London: Routledge. pp. 87–104.

18. Hawkins, G. (2013). Made to be wasted: PET and the topologies of disposability. In Gabrys, J., Hawkins, G. and Michael, M. (Eds.) *Accumulation: The Material Politics of Plastic*. London: Routledge. pp. 49–67.

19. London, B. (1932). Ending the depression through planned obsolescence. Original pamphlet, https://web.archive.org/web/20120819154515/http://upload.wikimedia.org/wikipedia/commons/2/27/London_%281932%29_Ending_the_depression_through_planned_obsolescence.pdf.

20. Barnes, D. K. A., Galgani, F., Thompson, R. C. and Barlaz. (2009). *Philosophical Transactions of the Royal Society B*. 364, 1985–1998. Doi: 10.1098/rstb.2008.0205.

21. Gabrys, J. (2013). Plastic and the work of the biodegradable. In Gabrys, J., Hawkins, G. and Michael, M. (Eds.) *Accumulation: The Material Politics of Plastic*. London: Routledge. pp. 208–227.

22. Takada, S. (2013). International pellet watch: Studies of the magnitude and spatial variation of chemical risks associated with environmental plastics. In Gabrys, J., Hawkins, G. and Michael, M. (Eds.) *Accumulation: The Material Politics of Plastic*. London: Routledge. pp. 184–207.

23. Cozar, A. *et al.* (2014). Plastic debris in the open ocean. *Proceedings of the National Academy of Science*. 111(128), 10239–10244; Jambeck, G., Wilcox, S., Perryman, A. and Narayan, L. (2015). Plastic waste inputs from land into the ocean. *Science*, (347), p. 768–771; Geyer, R., Jambeck, J. R. and Lavender, K. L. (2017). Production, use, and fate of all plastics ever made. *Science Advances*, 3(7), e1700782.

24. Crutzen, P. and Stoermer, E. (2000). The 'Anthropocene'. *The International Geosphere-Biosphere Programme Newsletters*. 41, 17–18; Bonneuil, C. and Fressoz, J. B. (2015). *The Shock of the Anthropocene*. New York: Verso Books.

25. Zalasiewicz, J. *et al.* (2016, March). The geological cycle of plastics and their use as a stratigraphic indicator of the Anthropocene. *Anthropocene*. 13, 4–17.

26. Proctor, R. and Schiebinger, L. (Eds.) (2008). *Agnotology the Making and Unmaking of Ignorance*. Palo Alto: Stanford University Press.

27. Valdemar, d Ambrières. (2019). Plastics recycling worldwide: Current overview and desirable changes. *Field Actions Science Reports*. 9, 12–21.

Chapter 7

Silicon

M. Grant Norton

*The Honors College and School of Mechanical
and Materials Engineering,
Washington State University, Pullman WA 99164, USA*

1. Introduction

In 1966, 481 million silicon transistors were sold [1]. Up until that year germanium was the dominant transistor material, but once that position was relinquished the importance of silicon was on an exponential growth trajectory. We had entered the silicon age. Almost every aspect of our lives was to become influenced by how electrons moved through microscopic regions of silicon a fraction of the size of a red blood cell. With transistors now cheaper than a single grain of rice, the technology enabled by silicon integrated circuits (silicon "chips") has become universally available [2].

2. The Silicon *p–n* Junction

The *p–n* junction in silicon — the basic building block of a silicon transistor — was discovered and named by Russell Ohl and Jack Scaff, both scientists at Bell Laboratories ("Bell Labs") in Holmdel, New Jersey, in 1940 [3]. The *n*-type (or negative) region was characterized by an excess of electrons. The *p*-type (or positive) region was characterized by a deficiency of electrons (an excess of positively charged holes). Where they met was the *p–n* junction. Subsequent experiments showed that *n*-type conductivity in silicon could be increased by adding phosphorus — an element from Group 5 of the Periodic Table. Small amounts of boron — an element from Group 3 — were found to increase *p*-type conductivity [4]. At the junction there was recombination of the excess charge carriers forming a depletion zone. By controlling the bias applied to the *p–n* junction, it was possible to control the movement of free charge carriers on either side of the junction. Current can flow easily in one direction, whereas in the other direction it is limited by a leakage current, which is generally very small. This behavior is known as rectification and is the characteristic property of a diode.

An early use for these junctions was as microwave radar detectors, an important application with the United States soon to join World War II following the Japanese attack on Pearl Harbor in December 1941. Millions of diodes comprising *p–n* junctions formed on tiny crystal slivers of silicon or germanium were fabricated for use in Allied radar receivers. G.L. Pearson and B. Sawyer, both at the Bell Labs, Murray Hill, New Jersey location, made the prototype alloy silicon diode, which was 1,000 times smaller than its germanium predecessor [5]. Not only was the silicon diode smaller, it had superior properties to the germanium diode largely because of the greater band gap energy, 1.11eV compared to 0.66eV.

3. The Silicon Chip

Since the invention of the point-contact transistor in 1947 by John Bardeen and Walter Brattain all transistors had been made with germanium [6]. That situation changed in 1954 when Bell Labs' chemist Morris Tanenbaum fabricated the first silicon transistor [7]. Just three months later a research team at Texas Instruments made a successful silicon transistor, also an *n–p–n* structure. On May 10, 1954, Gordon Teal who had left Bell Labs for Texas Instruments announced at the Institute of Radio Engineers National Conference on Airborne Electronics in Dayton, Ohio, that silicon transistors were commercially available. At the same meeting Teal demonstrated a significant advantage that silicon transistors had over their germanium counterpart — they worked when hot. Three years after Teal's presentation, germanium transistors were still outselling the considerably more expensive silicon units by 25 to 1. In 1957, the average cost of a germanium transistor was $1.85 compared to $17.81 for silicon [8].

An important property of silicon, in addition to its band gap energy, is that when heated in air or steam it oxidizes to form a dense, protective dielectric surface layer of amorphous silicon dioxide (SiO_2). This observation was made in 1955 during a serendipitous accident when a fire inside a diffusion furnace caused water vapor to leak into the high-temperature chamber where it oxidized the surface of the silicon wafer. It was later found that certain dopants, for instance, gallium, could diffuse through the oxide layer while others including boron and phosphorus could not. By opening small windows into the oxide layer, impurities that could not diffuse through the oxide layer could be introduced to selectively dope specific regions of the wafer [9, 10]. Intricate patterning of the windows was possible using photolithography, a technique that was already used to pattern printed circuit boards. The oxide layer formed on the surface of silicon was critical to the large-scale manufacturing of silicon integrated circuits and is a unique property of silicon that is not shared among other elemental and compound semiconductors [11].

In May 1958, Jack Kilby had recently joined Texas Instruments and was working on miniaturizing electronic circuits, which contained combinations of resistors, capacitors and diodes all made of different materials. These components were then laboriously hand assembled into the final device. Kilby wondered whether it was possible to make all the devices from a single material and integrate them to form a completed circuit. By September, Kilby had constructed a circuit using slivers of doped germanium etched to form *p–n–p* transistor, capacitor and resistor elements. These elements were then connected using fine "flying" gold wires to produce a phase-shift oscillator — a device that outputs a sine wave shifted to higher frequency. The invention of "miniaturized electronic circuits" was filed as a United States Patent in February 1959 [12]. Included among the suitable substrates were the elemental semiconductors germanium and silicon and the compound semiconductors gallium arsenide, aluminum antimonide and indium antimonide. An essential aspect described in Kilby's patent was forming an insulating oxide layer on the semiconductor, which is then selectively removed depending on the required function of the circuit.

A critically important development following from Kilby's work was the invention of the planar manufacturing process by Jean Hoerni, a physicist at Fairchild Semiconductor Corporation [13]. The process creates doped layers below one face of a silicon wafer, leaving the oxide layer in place to protect the underlying circuitry. Furthermore, Hoerni's invention demonstrated the ability to form many transistors on a single silicon wafer. The planar transistor, number 2N1613, was commercially introduced by Fairchild Semiconductor in 1960.

The next logical step was to combine Hoerni's planar process for the transistor with Kilby's component integration to produce a monolithic integrated circuit. This was the idea of Robert Noyce, a co-founder of Fairchild Semiconductor. Noyce's patent was filed in July 1959 and described how diodes, transistors and capacitors could be created in silicon by diffusion and then interconnected with aluminum lines formed on top of the protective silica layer; an entire electrical circuit on a single silicon chip [14].

Led by Jay Last, Fairchild Semiconductor fabricated the first planar integrated circuit in 1960 in what became known as Silicon Valley — the area of California between Palo Alto and San Jose. The planar integrated circuit, and the method to produce it, has been the basis for the electronics industry for the past 60 years and looks very much like it will continue into the foreseeable future.

4. A Fortunate Accident

An essential requirement in producing silicon for integrated circuits is the need to have very large single crystals. The majority of silicon single crystals are made using a crystal pulling technique that was first demonstrated, accidentally, in 1913 by Polish scientist Jan Czochralski. While writing up his notes, Czochralski dipped his pen into a crucible containing molten tin at a temperature of 230°C, rather than the nearby ink pot. Realizing his mistake Czochralski quickly withdrew the pen, which had a trail of solidified tin hanging from the nib. He had discovered a process of crystallization by pulling from the surface of a melt. The paper describing the crystal growth technique, which is frequently referred to simply as the Cz method, was subsequently published in 1917 [15]. Gordon Teal and

John Little of Bell Labs demonstrated the ability of a modified Cz process for semiconductor crystals by growing a single crystal of germanium by slowly withdrawing a seed crystal from a melt of very pure germanium (melting temperature 938°C) [16]. At the meeting of the American Physical Society in Washington D.C in May 1952, Teal and Ernest Buehler reported the successful growth of large single crystals of silicon by pulling from the melt (melting temperature 1414°C) and a patent was filed for a process for producing semiconductive crystals of uniform resistivity in June 1951 [17, 18].

To grow silicon crystals by the Cz process, a silicon seed crystal of the desired orientation is lowered into a silica crucible containing the melt at a temperature about 1420°C. The melt temperature is reduced slightly until the silicon begins to solidify onto the seed. Once this happens, the seed is slowly raised allowing more of the melt to solidify into a growing crystal. It is typical for the seed to be rotated to even out any thermal inhomogeneities in the surrounding furnace. A typical growth rate is 100–200 mm/h with a rotation rate of 10–20 rpm. To avoid excessive erosion of the crucible, it can also be rotated at the same speed as the seed.

Because silicon is highly reactive with oxygen, the entire growth process must be carried out in an inert, argon atmosphere. The melt will contain small amounts of the appropriate dopant — typically phosphorus or boron to produce boules that are either *n*-type or *p*-type, respectively. A heavily doped silicon melt might have $> 10^{19}$ B atoms/cm^3 [19].

Figure 1 shows the seed end of a Cz silicon boule with the very narrow neck region. (The other end is the tang.) The striations — growth rings — parallel to the crystal-growth interface are commonly seen in Cz grown silicon and can arise from a number of causes; changes in the growth rate caused by variations in pulling speed, segregation of impurities and temperature fluctuations at the solid–liquid interface.

In addition to requiring single crystals of high purity with the desired dopant concentration, it is important that the crystals are dislocation-free. Dislocations are non-equilibrium defects, which means that unlike point defects they can be entirely removed from a material permitting dislocation-free crystals to be grown under controlled conditions [20]. Although for many materials even large amounts of dislocations do not present any problems, in fact in metals they are often desirable, in a semiconductor

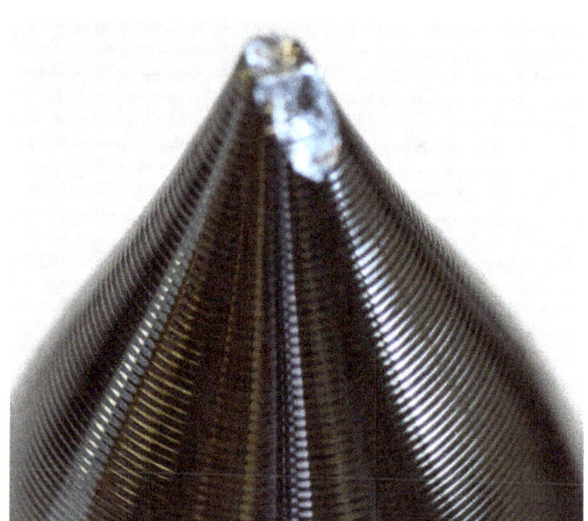

Figure 1.   Growth rings formed at the seed end of a Cz silicon boule. (C.B. Carter and M.G. Norton, *Ceramic Materials: Science and Engineering*, 2nd edition, New York: Springer, 2013, p.529. Republished here under Springer Copyright Transfer Statement.)

they must be avoided. A. G. Chynoweth and G. L. Pearson of Bell Labs found that breakdown in silicon *p-n* junctions occurred preferentially at dislocations and dislocations were shown to act as recombination centers [21, 22]. Sir Charles Frank and William Thornton Read had already shown how dislocations in silicon could multiply under an applied stress [23]. A dislocation pinned at two points could bow out under an applied stress until the two segments meet and annihilate forming a complete loop together with the original pinned dislocations. The loop expands and the process continues creating a succession of dislocation loops.

William Dash a crystal grower at General Electric Research Laboratories in Schenectady, New York, modified the Cz process to produce silicon crystals completely free of dislocations [24]. During the initial stages of growth Dash reduced the diameter of the neck region allowing dislocations to glide to the surface of the hot crystal and disappear. For silicon with the diamond-cubic structure, dislocations glide on the {111} planes, which make a large angle with the growth direction, which is typically <111> or <100>. Once all the dislocations have been annihilated, the growth rate is slowed until a boule of the necessary diameter is produced. Typically, the minimum size for the neck region is 3 mm diameter and 30 mm length. This narrow rod can support a boule weight of about 200 kg. About 95% of all single crystal silicon is produced by the Cz method. The remainder is mainly grown using the float zone (FZ) method [25].

With large, appropriately doped, dislocation-free silicon crystals, it was possible to mass produce wafers, now up to 18 inches in diameter, as a substrate for the fabrication of billions of transistors.

5. Origin and Extraction

Before we describe another critical electronic application for silicon, solar cells, let's look at the origin of the element, its abundance, a brief history and how silicon is extracted and processed into a useable form. Silicon was formed after the Big Bang in the explosions of massive stars and ultra-dense white dwarfs. The element's reactivity with oxygen means that silicon is not found in its native state in the Earth's crust, but rather bound with oxygen in the form of oxide minerals including quartzite (silica), clays and feldspars. Silicon is the second most abundant element in the crust, accounting for about 27% by weight. Only oxygen is present in a greater amount. Worldwide annual production of silicon is approximately 7 million metric tons [26]. The largest producers are China, Russia, Norway and Brazil. The main consumers are the aluminum alloy producers, the chemical industry and the manufacturers of chips for computers and photovoltaic cells.

The discovery of silicon is credited to Swedish scientist Jöns Jacob Berzelius. In 1824, Berzelius prepared a pure form of amorphous silicon by heating potassium with fluorosilic acid gas. Although the product of the reaction was contaminated with potassium silicide, Berzelius was able to remove it by reacting with water, leaving a relatively pure silicon powder. It was not until 1854 when crystalline silicon was prepared by French chemist Henri Etienne Sainte-Claire Deville.

The existence of the element had already been predicted as early as 1787 and given the name *silicium* by English chemist Sir Humphry Davy in 1808. The name was derived from the Latin "silicis" meaning flint, which is a silicon-containing sedimentary rock, with the ending "ium" added because Davy thought the element would be metallic and followed the accepted naming convention in place since 1800. But silicon showed "not the smallest evidence for its metallic nature" and was renamed by Sottish chemist and mineralogist Thomas Thomson [27].

Silicon is isostructural with diamond, whose crystal structure had been determined in 1913 by William Henry Bragg and William Lawrence Bragg using the then recently developed technique of X-ray diffraction [28]. The Bragg's subsequently were awarded the Nobel Prize in Physics "for their services in the analysis of crystal structure by means of X-rays."

Before large silicon crystals can be pulled using the Cz process, silicon must be extracted from its ore and converted into a highly pure raw material. The primary sources of silicon are silica and other silicon-containing oxide minerals. Metallurgical grade silicon with a purity of about 98% is produced by carbothermal reduction of the oxide at 1500°C using coal, coke or wood chips. Further purification requires converting the metallurgical grade silicon into either trichlorosilane ($SiHCl_3$) or silane (SiH_4) with exceptionally high levels of purity. The gas is then decomposed at high temperature by a chemical vapor deposition (CVD) process that takes place on the surface of specially prepared single crystal rods of silicon. This process is known in the industry as the Siemens method after its corporate inventors [29, 30]. The polycrystalline electronic grade (also known as semiconductor grade) silicon grows out from the surface of the rod producing a feather-like structure consisting of a central shaft with a series of branches and sub-branches [31]. The final polysilicon layer is several centimeters thick.

To create the temperatures high enough for the decomposition reaction (for silane the decomposition temperature is 850°C; it is above 1000°C for trichlorosilane) an electric current is passed through the silicon rod, which is shaped like a large upside-down U. This method requires expensive capital equipment, typically has high operating costs and often results in low silicon yields (15–30% in practice). The need for large amounts of electrical power is one reason why many silicon manufacturers are located near hydroelectric plants such as Grand Coulee Dam in central Washington state and the Liujiaxia Dam in China's Gansu Province.

The formation of silicon by the CVD reaction of chlorosilanes onto hot surfaces significantly predates the work of Siemens engineers. In the early 1890s, Rudolf Langhans of Berlin, Germany, used the method in an approach to make a more reliable filament for the recently patented incandescent lamp [32]. Langhans's substrate was not high-purity silicon, it was the electrically conductive silicon carbide. For a short period from 1891 to 1893, Langhan worked on an unsuccessful silicon-impregnated cellulose filament.

6. Solar Silicon

Towering above the sloping semi-arid plateau and west of the sign boasting Grant County, Washington, as "The nation's leading potato producing county" stand the fluidized bed reactors (FBR) of Renewable Energy Corporation (REC) in Moses Lake. In these reactors, polysilicon is produced in a granular form for the photovoltaic industry; which now consumes more silicon than that used for producing silicon chips [33].

The photovoltaic effect — the principle behind how solar cells operate — was discovered by Russell Ohl of Bell Labs in 1940 when he observed that light striking a p–n junction caused an electric current by electrons flowing from the n-side of the junction to the p-side [34]. Using large p–n junctions formed by diffusing boron atoms into n-type silicon wafers, Calvin Fuller, Daryl Chapin and Gerald Pearson made the first silicon solar cells. The cells had an energy conversion efficiency of a modest 6%, but by the end of the 1950s solar cells were providing power for terrestrial applications including rural telephone systems and extraterrestrial applications such as space satellites.

The FBR process for production of polysilicon was developed, at least in part, because of the demand from the solar cell industry for a lower cost and less energy-intensive product [35, 36]. Fluidized bed reactors produce granular silicon by pyrolytic decomposition of silane-containing gases, typically trichlorosilane or silane. The process begins with silicon seed particles ranging from 100 to 2000 μm in diameter that are fluidized with a carrier gas such as hydrogen. An important and significant aspect of the FBR is the temperature range over which silicon can be produced. Gas decomposition can be achieved at temperatures as low as 600°C, as

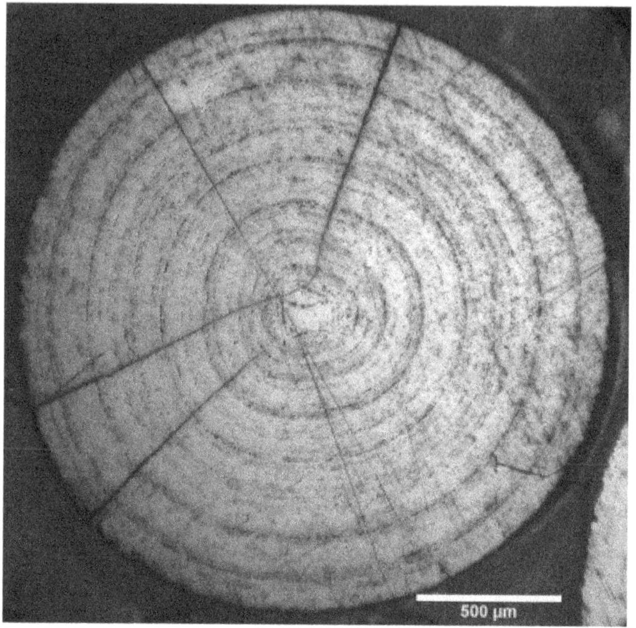

Figure 2.   Optical micrograph of an etched silicon granule from a FBR. The microstructure of the bead was revealed by etching with an isotropic HNA solution composed of 8% hydrofluoric acid, 25% nitric acid, and 67% acetic acid. (Reproduced from: Dahl MM, Bellou A, Bahr DF, Norton MG, Osborne EW. Microstructure and grain growth of polycrystalline silicon grown in fluidized bed reactors. *Journal of Crystal Growth* 2009; 311: 1496–1500 with permission from Elsevier.)

opposed to temperatures up to 1200°C that are used to produce bulk polysilicon via the Siemens process.

Once a desired particle size range has been achieved within the reactor, the granules fall to the bottom where they are collected, without disrupting the growth of the rest of the material. The continuous nature of the FBR process is another benefit compared to the batch-style Siemens approach.

Figure 2 shows an etched cross-section sample of a silicon granule collected from an FBR. Immediately apparent in the image is the series of concentric rings, "growth rings", caused by porosity differences within the granule as it moves between hotter and colder regions of the vertical reactor. Existing FBR plants can generate over 13,500 tons of polysilicon annually.

Solar photovoltaic capacity has doubled every two years over the past two decades. In the last 10 years alone, the United States has seen almost a 50% average annual growth rate in installed solar capacity. There are now more than 81 gigawatts (GW) of solar capacity installed nationwide, enough to power 15.7 million homes. Worldwide cumulative installation capacity at the end of 2019 exceeded 580 GW [37]. At the same time as capacity has increased, the cost of silicon photovoltaic cells has significantly decreased, dropping by more than 70% over the last decade [38]. Crystalline silicon solar cells continue to dominate the global photovoltaic market with about a 95% market share and offer a highest cell efficiency of 26.6% for single crystalline cells and 22.3% for multi-crystalline wafer-based technology [39, 40].

The impressive growth in the deployment of silicon photovoltaics has followed a trend set for increasing levels of integration in silicon chips. In 1965, Gordon Moore, Director of Research and Development at Fairchild Semiconductor, predicted that the number of components such as transistors that can fit on a silicon chip will double every year [41]. In 1965, that number was 60. A decade later the number of integrated components had shot up to over 60,000. Between 1971 and 2016 transistor counts have doubled every two years rather than every year — still an incredible technological achievement. The 1 million transistor mark was passed in 1989 and now devices exceeding one billion transistors are common.

7. Nanosilicon

Although silicon and carbon are in the same group in the Periodic Table, they do not share the same rich variety of allotropes. Carbon exists in many allotropic forms including the familiar diamond and graphite, but also as the soccer ball-shaped buckminsterfullerene and as single atom thick graphene sheets. The only allotropes of silicon are the diamond cubic crystal form and amorphous silicon. But the two elements do share some of the larger scale nanostructures. For instance, silicon has been formed into nanodots/quantum dots (QDs), nanowires (NWs), nanorods and nanoribbons [42–48]. In these geometries new applications have been found for silicon. As just a single example, luminescent silicon QDs have been used for imaging of pancreatic cancer cells [49]. Figure 3 (left) shows a

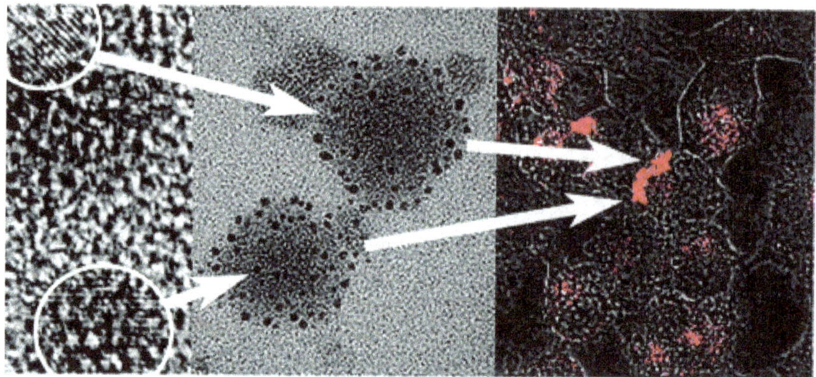

Figure 3. Composite image showing on the left individual crystalline silicon quantum dots about 4 nm in diameter. The center image shows a plurality of micelle-encapsulated quantum dots. The image on the right shows the overlap of the TEM image with the luminescence image showing uptake of the quantum dots by pancreatic cancer cells. (Reprinted with permission from: Erogbogbo F, Yong KT, Roy I, Xu GX, Prasad PN, Swihart MT. Biocompatible luminescent silicon quantum dots for imaging of cancer cells. *ACS Nano.* 2008; 2: 873–878. Copyright 2008 American Chemical Society.)

high-resolution transmission electron microscope (TEM) image of silicon QDs formed by laser pyrolysis of silane in an aerosol reactor followed by chemical etching in mixed hydrofluoric/nitric acid to reduce the particle size to around 4 nm in diameter. The QDs were then surface functionalized followed by micellar encapsulation to increase hydrophilicity. Within each micelle there were multiple QDs as shown in Figure 3 (middle). The combined TEM/luminescence image, Figure 3 (right), shows that the micelle encapsulated QDs were preferentially taken up by pancreatic cancer cells *in vitro*. The results suggest that silicon QDs have the potential for use as a nontoxic optical probe for biomedical diagnostics [49].

While advances in the processing of silicon integrated circuits have allowed transistor counts to keep pace with the predictions of Moore's law, will silicon be able to maintain its dominance indefinitely? The smallest transistors in production are now only 5 nm. Intel has plans that by 2029 transistors will shrink to a miniscule 1.4 nm, equivalent to only 12 silicon atoms across. While some researchers are studying approaches to make increasingly small transistors, other groups are exploring the use of other materials as possible replacements for silicon. One such example is carbon nanotubes (CNTs).

Cees Dekker and colleagues at Delft University of Technology made the first CNT transistor, which formed the basic building block of the CNT computer made in 2013 by Subhasish Mitra and Philip Wong of Stanford University, not far from where the first planar integrated circuits were made [50, 51].

The reason for exploring the use of CNTs is that, in principle, they allow faster and more energy efficient transistors, but overall CNT computers are much slower than their current silicon-based counterparts. A major challenge with CNTs and many of the other nanostructured forms of carbon is scaling up production of the material while ensuring product uniformity and consistency. It is also not clear with nanostructured transistors how the benefits of integration that have proved so successful for silicon over the past 55 years will carry over. The prototype CNT computer contained less than 200 transistors compared to the billions in a state-of-the-art silicon chip.

Figure 4 shows an example of a silicon NW field effect transistor (FET) [52]. The arrangement consists of a silicon core about 20 nm wide terminated at each end by nickel silicide forming a pair of Schottky junctions. (A Schottky junction, or Schottky diode, is the junction formed by a semiconductor with a metal.) The entire axial heterostructure is embedded within an amorphous SiO_2 shell. Charge carrier injection through each Schottky junction is controlled independently by two separate top metal gate electrodes. The individual NWs were fabricated via the vapor–liquid–solid (VLS) mechanism, a widely used approach for the synthesis of one-dimensional nanostructures [53]. Such an NW transistor, like related CNT transistors, cannot benefit from the process integration that has been achieved with planar transistors. However, an advantage of the silicon NW FET is that it uses intrinsic silicon, it does not require doping.

Silicon has been without question the enabling material for the current generation of computers and the associated electronic technology. Silicon may also be the material at the heart of next generation quantum computers [54]. Keiji Ono of the Advanced Device Laboratory of RIKEN in Japan explains: "… we are attempting to develop a quantum computer based on the silicon manufacturing techniques currently used to make computers and smartphones. The advantage of this approach is that it can leverage existing industrial knowledge and technology" [55]. There are a number of different silicon architectures that are being explored for

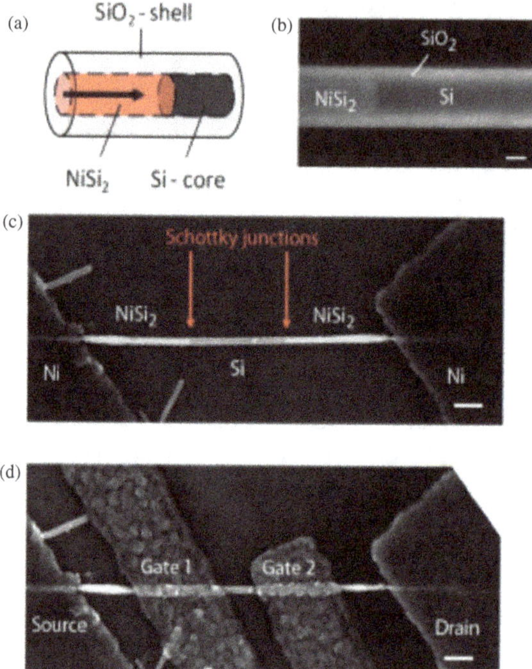

Figure 4.   Nanowire heterostructure formation and SEM image of a reconfigurable NW FET. (a) Schematic of axial Ni silicidation within the SiO$_2$ shell. (b) SEM of the formed interfaces of NiSi$_2$, Si, and SiO$_2$. (c) SEM of NW FET prior to gate electrode patterning. The red arrows indicate the location of the Schottky junctions. (d) Nanowire FET with top gate electrodes, gates 1 and 2, overlapping the Schottky junctions. The scale bar is 20 nm in (b) and 200 nm in (c,d). (Reprinted with permission from: Heinzig A, Slesazeck S, Kreupl F, Mikolajick T, Weber WM. Reconfigurable silicon nanowire transistors. *Nano Letters*. 2012; 12: 119–124. Copyright 2012 American Chemical Society.)

quantum computing. One example is shown in Figure 5 [56]. It consists of a ^{28}Si substrate with atomically straight step-edges that have been formed by heat treatment of an intentionally off-axis (vicinal) (111)-oriented silicon wafer. After the substrate has been prepared, ^{29}Si isotopes are deposited under ultra-high vacuum conditions and align preferentially along the step edges forming an atomically straight nanowire. These wires will be used for quantum computing.

Although quantum computers are in their relative infancy, first being proposed in 1980, they are predicted to follow an exponential growth in

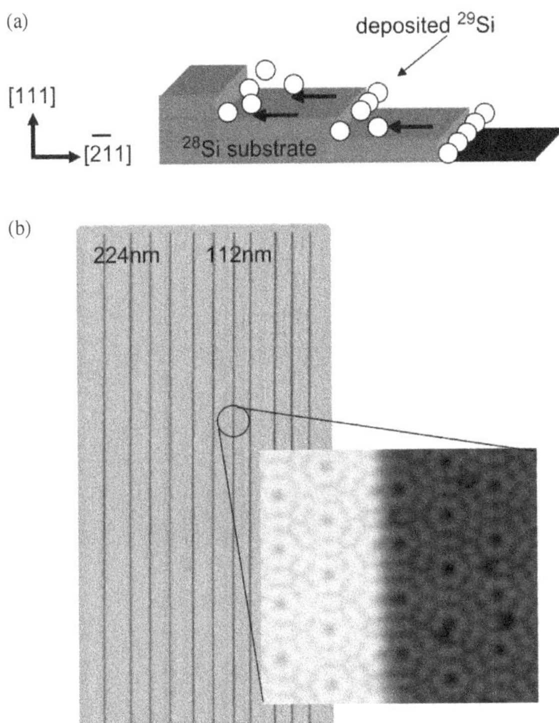

Figure 5. A schematic of single ^{29}Si wire formation at the straight step edge of the vicinal ^{28}Si (111) surface in the step-flow mode of epitaxial growth. (B) A STM picture of the straight step edge vicinal Si (111) surface. The right figure is a close up of the step-edge structure. (Reproduced from: Itoh KM. An all-silicon linear chain NMR quantum computer. *Solid State Communications*. 2005; 133: 747–752 with permission from Elsevier.)

computing power [57]. Rose's law is a quantum computing version of Moore's law [58].

Each day billions of people connect to the Internet through an estimated 30 billion devices. These devices allow us to keep in touch with friends and family, to share information, to keep us safe and healthy, and to remotely control many aspects of our daily lives. All of these innovations utilize silicon integrated circuits. The Internet of Things and artificial intelligence are currently possible because of our ability to produce billions of tiny transistors on single pieces of silicon. Silicon is also proving itself to be important in the form of various nanostructures, which are opening up new application areas and technologies. Even next generation quantum computers may find that they too are enabled by silicon.

References

1. *Electronic Industries Yearbook.* (1967). Electronic Industries Association, Marketing Services Department.
2. Qualman, D. (2015). Data shows the enormous increase in transistor production per capita since 1955, http://www.darrinqualman.com/global-production-transistors/.
3. Scaff, J. H. and Ohl, R. S. (1947). Development of silicon crystal rectifiers for microwave radar receivers. *Bell System Technical Journal.* 26, 1–30.
4. Scaff, J. H., Theuerer, H. C. and Schumacher, E. E. (1949). P-type and N-type silicon and the formation of the photovoltaic barrier in silicon ingots. *Transactions of the Metallurgical Society of AIME.* 185, 383–388.
5. Pearson, G. L. and Sawyer, B. (1952). Silicon *P-N* junction alloy diodes. *Proceedings of the IRE.* 40, 1348–1351.
6. Bardeen, J. and Brattain, W. (1948). The transistor, a semi-conductor triode. *Physical Review.* 74, 230–231.
7. Tanenbaum, M., Valdes, L. B., Buehler, E. and Hannay, N. B. (1955). Silicon *n-p-n* grown junction transistors. *Journal of Applied Physics.* 26, 686–692.
8. Morris, P. R. (1990). *A History of the World Semiconductor Industry.* London: Peter Peregrinus. p. 49.
9. Derick, L. and Frosch, C. J. (1957). Oxidation of semiconductive surfaces for controlled diffusion. United States Patent 2,802,760.
10. Frosch, C. J. and Derick, L. (1957). Surface protection and selective masking during diffusion in silicon. *Journal of the Electrochemical Society.* 104, 547–552.
11. Gregory, O. J., Pruitt, L. A., Crisman, E. E, Roberts, C. and Stiles, P. J. (1988). Native oxides formed on single-crystal germanium by wet chemical reactions. *Journal of the Electrochemical Society.* 135, 923–929.
12. Kilby, J. S. (1964). Miniaturized electronic circuits. United States Patent 3,138,743.
13. Hoerni, J. A. (1962). Method of manufacturing semiconductor devices. United States Patent 3,025,589.
14. Noyce, R. N. (1961). Semiconductor device-and-lead structure. United States Patent 2,981,877.

15. Czochralski, J. (1917). A new method for the measurement of the velocity of crystallization of metals. *Zeitschrift für Physikalische Chemie.* 92, 219–221.
16. Teal, G. K. and Little, J. B. (1950). Growth of germanium single crystals. *Physical Review.* 78, 647.
17. Teal, G. K. Buehler E. (1952). Growth of silicon single crystals and of single crystal silicon *p-n* junctions. *Physical Review.* 87, 190.
18. Buehler, E. and Teal, G. K. (1956). Process for producing semiconductive crystals of uniform resistivity. United States Patent 2,768,914.
19. Hoshikawa, K., Huang, X., Taishi, T., Kajigaya, T. and Iino, T. (1999). Dislocation-free Czochralski silicon crystal growth without the dislocation-elimination-necking process. *Japanese Journal of Applied Physics.* 38, L1369.
20. Frenkel, Y. (1926). Uber die Wärmebewegung in festen und flüssigen Körpern. *Zeitschrift für Physik.* 35, 652–669.
21. Chynoweth, A. G. and Pearson, G. L. (1958). Effect of dislocations on break-down in silicon *p-n* junctions. *Journal of Applied Physics.* 29, 1103.
22. Bemski, G. (1958). Recombination in semiconductors. *Proceedings of the IRE.* 46, 990–1004.
23. Frank, F. C. and Read, Jr W. T. (1950). Multiplication processes for slow moving dislocations. *Physical Review.* 79, 722.
24. Dash, W. C. (1959). Growth of silicon crystals free from dislocations. *Journal of Applied Physics.* 30, 459–474.
25. Shimura, F. (2017). Single-crystal silicon: Growth and properties. In Kasap, S. and Capper, P. (Eds.) *Springer Handbook of Electronic and Photonic Materials.* Cham: Springer.
26. United States Geological Survey (USGS). Annual mineral summary, https://pubs.usgs.gov/periodicals/mcs2020/mcs2020-silicon.pdf.
27. Thomson, T. (1817). *A System of Chemistry in Four Volumes.* 5th Ed., Vol. 1. London: Baldwin, Cradock and Joy. p. 252.
28. Bragg, W. H. and Bragg, W. L. (1913). The structure of the diamond. *Proceedings of the Royal Society of London A.* 89, 277–291.
29. Sirtl, E. and Seiter, H. (1969). Smooth surfaced polycrystals. United States Patent 3,476,640.
30. Reuschel, K. (1978). Method of depositing elemental amorphous silicon. United States Patent 4,068,020.
31. Fancher, R. W., Watkins, C. M., Norton, M. G., Bahr, D. F. and Osborne, E. W. (2001). Grain growth and mechanical properties in bulk polycrystalline silicon. *Journal of Materials Science.* 36, 5441–5446.

32. Langhans, R. (1890). Improved filament for incandescent lamp. United States Patent 420,881.

33. Zbib, M. B., Tarun, M. C., Norton, M. G., Bahr, D. F., Nair, R., Randall, N. X. and Osborne, E. W. (2010). Mechanical properties of polycrystalline silicon solar cell feed stock grown via fluidized bed reactors. *Journal of Materials Science*. 45, 1560–1566.

34. Ohl, R. S. (1946). Light-sensitive electric device. U.S. Patent 2,402,662.

35. Dahl, M. M., Bellou, A., Bahr, D. F., Norton, M. G. and Osborne, E. W. (2009). Microstructure and grain growth of polycrystalline silicon grown in fluidized bed reactors. *Journal of Crystal Growth*. 311, 1496–1500.

36. Osborne, E. W., Spangler, M. V., Allen, L. C., Geertsen, R. J., Ege, P. E., Stupin, W. J. and Zeininger, G. (2011). Fluid bed reactor. United States Patent 8,075,692 B2.

37. Photovoltaics Report. (2020). Fraunhofer Institute for Solar Energy Systems with support of PSE Projects GmbH, https://www.ise.fraunhofer.de/content/dam/ise/de/documents/publications/studies/Photovoltaics-Report.pdf.

38. Solar Energy Industries Association (SEIA). https://www.seia.org/solar-industry-research-data.

39. Wang, Z., Zhu, X., Zuo, S., Chen, M., Zhang, C., Wang, C., Ren, X., Yang, Z., Liu, Z., Xu, X., Chang, Q., Yang, S., Meng, F., Liu, Z., Yuan, N., Ding, H., Liu, S. and Yang, D. (2020). 27%-efficiency four-terminal perovskite/silicon tandem solar cells by sandwiched gold nanomesh. *Advanced Functional Materials*. 30, 1908298.

40. Green, M. A., Dunlop, E. D., Hohl-Ebinger, J., Yoshita, M., Kopidakis, N., Ho-Baillie, A. W. Y. (2020). Solar cell efficiency tables (Version 55). *Progress in Photovoltaics: Research and Applications*. 28, 3–15.

41. Moore, G. (1965). Cramming more components onto integrated circuits. *Electronics Magazine*. 38(8), 114.

42. Holmes, J. D., Ziegler, K. J., Doty, R. C., Pell, L. E., Johnston, K. P. and Korgel, B. A. (2001). Highly luminescent silicon nanocrystals with discrete optical transitions. *Journal of the American Chemical Society*. 123, 3743–3748.

43. Cavarroc, M., Mikikian M, Perrier G, Boufendi L. Single-crystal silicon nanoparticles: An instability to check their synthesis. *Applied Physics Letters*. 2006; 89: 013107.

44. Ma, D. D. D, Lee, C. S., Au, F. C. K., Tong, S. Y. and Lee, S. T. (2003). Small-diameter silicon nanowire surfaces. *Science*. 299, 1874–1877.

45. Schmidt, V., Wittemann, J. V., Senz, S. and Gósele, U. (2009). Silicon nanowires: A review on aspects of their growth and their electrical properties. *Advanced Materials*. 21, 2681–2702.

46. Fan, J. G., Tang, X. J. and Zhao, Y. P. (2004). Water contact angles of vertically aligned Si nanorod arrays. *Nanotechnology*. 15, 501–504.

47. Zschech, D., Kim, D. H., Milenin, A. P., Scholz, R., Hillebrand, R., Hawker, C. J., Russell, T. P., Steinhart, M. and Gösele, U. (2007). Ordered arrays of <100>-oriented silicon nanorods by CMOS-compatible block copolymer lithography. *Nano Letters*. 7, 1516–1520.

48. Baca, A. J., Meitl, M. A., Ko, H. C., Mack, S., Kim, H. S., Dong, J., Ferreira, P. M. and Rogers, J. A. (2007). Printable single-crystal silicon micro/nanoscale ribbons, platelets and bars generated from bulk wafers. *Advanced Functional Materials*. 17, 3051–3062.

49. Erogbogbo, F., Yong, K. T., Roy, I., Xu, G. X., Prasad, P. N. and Swihart, M. T. (2008). Biocompatible luminescent silicon quantum dots for imaging of cancer cells. *ACS Nano*. 2, 873–878.

50. Trans, S. J., Verschueren, A. R. M. and Dekker, C. (1998). Room-temperature transistor based on a single carbon nanotube. *Nature*. 393, 49–52.

51. Shulaker, M. M., Hills, G., Patil, N., Wei, H., Chen, H. Y., Wong, H. S. P. and Mitra, S. (2013). Carbon nanotube computer. *Nature*. 501, 526–530.

52. Heinzig, A., Slesazeck, S., Kreupl, F., Mikolajick, T. and Weber, W. M. (2012). Reconfigurable silicon nanowire transistors. *Nano Letters*. 12, 119–124.

53. Carter, C. B. and Norton, M. G. (2013). *Ceramic Materials: Science and Engineering*. 2nd Ed. New York: Springer. p. 537.

54. Kane, B. E. (1998). A silicon-based nuclear spin quantum computer. *Nature*. 393, 133.

55. Ono, K., Giavaras, G., Tanamoto, T., Ohguro, T., Hu, X. and Nori, F. (2017). Hole spin resonance and spin-orbit coupling in a silicon metal-oxide-semiconductor field-effect transistor. *Physical Review Letters*. 119, 156802.

56. Itoh, K. M. (2005). An all-silicon linear chain NMR quantum computer. *Solid State Communications*. 133, 747–752.

57. Benioff, P. (1980). The computer as a physical system: a microscopic quantum mechanical Hamiltonian model of computers as represented by Turing machines. *Journal of Statistical Physics*. 22, 563–591.

58. Tanburn, R., Okada, E. and Dattani, N. (2015). Reducing multi-qubit interactions in adiabatic quantum computation without adding auxiliary qubits. Part 1: The "deduc-reduc" method and its application to quantum factorization of numbers. arXiv:1508.04816.

Chapter 8

Nanocarbons

Sacha Loeve

University Lyon 3 Jean Moulin, France

Graphite and diamond have been used since prehistory, long before they were recognized as two *allotropes* of elemental carbon [1, pp. 39–61]. From *allos*, "other" + *tropos* "manner", "way" or "mode", allotropy refers to the ability of an element to form different simple bodies made up of the same atoms bound together in various specific ways. Of all elements, carbon is the champion of allotropy. At the threshold of bulk matter, the nanometer scale,[1] carbon affords many more allotropes than its two worldly macroscopic forms graphite and diamond. These recent avatars of good old carbon have so unexpected behaviors that they figure out as star materials worthy of heavy Research and Development investments. Since 2000 they have generated great expectations in all technological sectors. As a few of them enter in industrial production, they are exemplars of the economy of promises attached to new materials.

1. A Zoo of Exotic Structures

The spherical structure of fullerene[2] came to be known in the 1980s as the "third carbon allotrope" [2, 3]. "Buckyballs" generated renewed interest

---

[1] One-billionth of a meter.

[2] Named after the visionary architect Buckminster Fuller, famous for his geodesic domes.

Figure 1.   A collection of carbon allotropes. From top left: diamond, graphite, lonsdaleite (hexagnoal diamond), amorphous carbon, fullerenes C_{60}, C_{70}, C_{960}, single and multi-walled carbon nanotubes, nanosausage, nanotori, nanobuds, nanohorns, nanoscrolls, carbon onion, nanocone, nanoribbon, Möbius carbon ribbon, nanofoam, carbon bamboos, fullerene pipes. Bottom: graphene, the mother structure of all hexagonal nanocarbons. Picture by the author (objects not on scale).

in two other carbon allotropes, nanotubes and graphene, that had been repeatedly observed and described long before they were 'discovered' in the 1990s and 2000s, respectively.

Fullerenes, nanotubes and graphene are the most iconic carbon nano-materials, because of their typical "low-dimensional" geometries. However they came along with a whole menagerie of other exotic nanocarbons, characterized by electron microscopy, diffraction, mass or X-ray spectroscopy. To mention only a few: nanocones, fullerene pipes, nano-horns, nanofoam, nanotori, nanoscrolls, nano-onions, nanobuds, nanorib-bons and even nanobamboos (Figure 1). This large family thrives on carbon's ability to form a wide array of two-dimensional sheet structures that can loop back on themselves, bend, fold or twist in many different ways, each one affording specific properties. The number of different nano-carbons can reasonably be considered indefinite. Assuming that low-dimensions underpin at least partially higher-ones, such unlimited form-generative potential bears something of a world-forming character.

2. Welcome to the Nanoworld

At the nanometer scale, carbon seems to open up a parallel world, the "nanoworld". Invisible to the naked eye, the nanoworld is nevertheless amply visible in popular media and science policy as it is saturated by promises. Nanocarbons are showcased as a cornucopia of futuristic prod-ucts: ultra-fast computers, ultra-thin sensors, touch-screens, pollutant-capture devices, targeted drug delivery carriers, not to mention a space elevator with a giant cable made of braided carbon nanotubes [4]! As Richard Smalley, the co-discoverer of fullerenes, once claimed, "These nanotubes are so beautiful that they must be useful for something."

The variety of potential nanocarbons depends not only on structural configurations (electron orbitals) but also on the process of synthesis (combustion, carbonization, graphitization, laser ablation, catalytic growth …).

Nanocarbons have changed the perspective on the old and familiar element carbon. The hero of the thermo-industrial revolution was still soaked in black with soot. The heroes of the nanoworld are pure surfaces, light, immaculate, geometric, with elegant structures not dissimilar to those generated in origami-kirigami games. How could a material so com-mon, used and overused by heavy industries, reveal such surprises? Nanocarbons have given new life and visibility to carbon chemistry. It is neither entirely organic nor inorganic chemistry, as nanocarbons are neither organic molecules nor crystalline solids. These new kinds of objects — carbon nanomaterials — have interconnected and reorganized entire research fields, henceforth labeled "nano"(-science, -technology, -engineering …). Like a phoenix, carbon was born again from its own ashes, embarking on new technological adventures.

The lifestory of nanocarbons could be told from different perspectives, depending on the people and contexts the biographer focuses on. Richard Smalley's story of the discovery of fullerenes, for example, differs from the one told by his co-discoverer Harold Kroto. The latter tells a story of dreamy and disinterested science, which begins by deciphering the sky, while the former insists on real-world problems and technological applications [5]. Carbon binds these stories together and could afford many others.

Another striking feature of the large family of nanocarbons is that there is no competition between them on the market. Rather they support each other: fullerenes helped to make nanotubes visible, which prompted a new interest in graphite, that led to the graphene. "The only character of true genius in the story is carbon," Smalley noted [3].

Without entering into the priority disputes raised by the discovery of carbon nanotubes [6], I will focus on "the true genius" itself. And I will follow the life trajectory of carbon nanotubes (CNTs) as exemplars of the large family of nanocarbons.

3. Mechanical Champions

In a sense, CNTs are the descendants of the carbon fibers manufactured for making composite materials for the past 50 years. Carbon fibers extended the market of plastics. Dubbed "the new steel", carbon fibers have a high elastic modulus and low density and rigidity that offers mechanical performance five or six times higher than aluminum or titanium alloys when used as reinforcing structures in composite materials. The processes for synthesizing CNTs are inherited from those of carbon fibers, such as electric arc furnaces, oxyacetylene torches and chemical vapor deposition techniques. The most used technique consists of cracking a molecular carbon-containing precursor (a plastic or a raw material like benzene) at the surface of small metallic catalytic nanoparticles: the molecule breaks down and releases carbon, which reorganizes itself to form a small cap (half a fullerene) in which carbon atoms bind together in hexagons and pentagons; the small cap grows gradually as carbon is incorporated and forms a nanotube. Depending on the synthesis conditions, the tubes can be single- or multi-walled, or a mixture of both. Carbon fibers are ultimately multi-wall CNTs.

It took years to uncover this type of growth. However, scientists were actually *rediscovering* a phenomenon that had long been observed in blast furnaces: the unexpected — and at that time undesirable — occurrence of "carbon filaments" so hard that they could crack bricks. These "filamentous carbons" were most likely multi-wall CNTs. And CNTs were also required in the formation process of carbon fibers, as the central structure around which the fiber thickens, as highlighted by the Japanese materials scientist Morinobu Endo.

In the 1970s, Endo unveiled the "central hollow tube" hidden in the core of his vapor-grown carbon fibers. But he was not interested in the tube itself. He viewed it as a key for the mechanical robustness of the fiber, its strongest part which "never breaks when the fiber breaks" [7]. The high mechanical performances of CNTs originate in the strength of the carbon–carbon bond in hexagonal planes, which only high temperatures can break. Very few materials have equivalent mechanical robustness combined with such flexibility. With Young modulus up to 1 teraPascal, a nanotube can be bent without breaking, and returned to its original state. Added to this is their extreme lightness: not only is carbon a light element, but they are hollow materials! Combining robustness, flexibility and lightness — everything that engineers dream of — CNTs have been touted as revolutionary materials.

However, these mechanical marvels are nothing new given their ascendancy with carbon fibers. Endo, for instance, was already massproducing these hollow microtubules for electric batteries when he learned — much to his stupefaction — that the same material had been 'discovered' by another Japanese scientist working at NEC Corporation, Sumio Iijima [8].

4. A Rising Star in Electronics

August 2019: A team from MIT assembled and operated a microprocessor made up of 14,000 carbon nanotube field-effect transistors (CNFET) using standard industrial methods [9]. How could CNTs become a star electronic material?

Twenty years ago, field-effect transistor based on a single nanotube was an unrealistic project: to isolate a nanotube, to place it on electrodes and measure its electric transport properties required an army of students. They would spend days and nights depositing a drop of CNT solution on a substrate and scanning the surface in the hope of finding a few carbon nanotubes well positioned on the top of two electrodes. However, these experiments confirmed the predictions based on calculus that there would be two types of carbon nanotubes: metallic and semiconductor. Graphite (at least very pure graphite) can be modeled as a stacking of hexagonal carbon monolayers, a pile of graphene sheets. Before graphene became a star material in the 2000s, the term "graphene" (coined after *graph*ite + benz*ene)* was used to refer to a single layer of graphite. This 2D structure was viewed as a laboratory curiosity. Graphene was promoted as a material in the 2000s when Andre Geim and Konstantin Novoselov crafted a tiny device showcasing its extreme electronic tunability [10]. In turn, the nanotube has been modeled as a cut-out of a single graphene sheet rolled on itself.

CNTs have a diameter between 0.4 to 100 nanometers for lengths that can go up to tens of centimeters. Like all "nano"-materials, their size matters, not for itself, but for the properties it brings out:

First, CNTs are both nano-objects and potentially macro-objects, and can be engineered at both scales. This is also true for an individual graphene sheet: sub-nano in thickness, it can be very large in the two dimensions.

Second, the length-to-diameter ratio, which can be as high as several millions-to-one and has no known theoretical limit, makes CNT properties extremely anisotropic, i.e. direction-dependent and tunable. As a result, this "one-dimensional conductor" affords "ballistic conduction"[3] and is highly versatile: a tiny alteration of the tube structure completely changes its electronic characteristics.

---

[3] In the inner vacuum space of the tube, electron transport shows negligible resistivity due to scattering and simply follows Newton's second law of motion.

Third, depending on their winding angle, diameter and length, the number of possible nanotube structures is theoretically indefinite.

Finally, because of the perfectly calculable matching between the rolling-up structure and the electronic properties of CNTs, between "fold and function", their potential for electronics appeared straightforward. CNTs were a serious candidate to move beyond silicon and continue the exponential growth in the miniaturization of electronic chips according to Moore's law.

5. From Laboratory to Industry

However, the difficulties in the synthetic process of CNTs moderated the enthusiasm. Not that they are difficult to make since carbon spontaneously assembles into fullerenes and nanotubes wherever it condenses. It has this "genius wired within it" as Smalley remarked in his Nobel Lecture. But CNTs are difficult to make in a predictable way that would deliver exclusively the specific tubes with the desired properties described by the theory. A "genius" is always somewhat whimsical!

Today, despite routine industrial processes, it is still a challenge to determine which synthesis conditions allow obtaining a tube of interest with the desired diameter, length and chirality. Even with more delicate techniques such as laser ablation or vaporization, their synthesis remains empirical and highly dependent on the environment, especially the catalyst. Some processes produce statistically more metallic CNTs, others more semiconductors. All of them yield mixtures of various purities, chiralities, diameters and lengths of CNTs, often agglomerated or bundled together like overcooked spaghetti (Figure 2). Up to now, the level of industrial mastery of the material's intrinsic properties remains largely below that which has been achieved with silicon for decades.

Industrial CNTs are rather used in common, incremental applications where mixtures of various purities can be used either to reinforce structures or as electric conductors. When immersed in the transparent film of solar panels, they can promote and preserve its transmittance. However, these mechanical and electrical applications are not exactly disruptive, they do not call for new technological concepts. In both cases, engineers

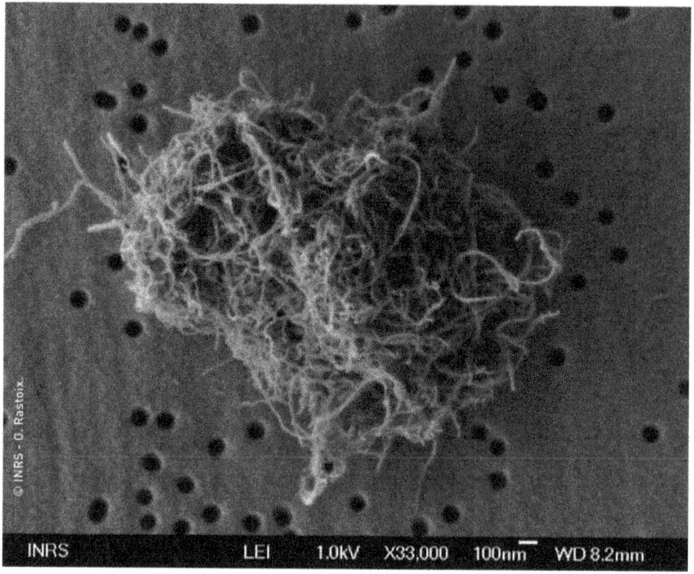

Figure 2.   Bundled nanotubes. Transmission electron microscopy [11].
Source: http://www.inrs.fr/media.html?refINRS=ND%202286.

use the morphology of the material and its ability to integrate with other well-controlled compounds. It just improves the existing technology.

Eventually, to enhance the yield of nanotubes with desired properties, one had to give up the idea of rational design and instead to improve post-synthesis sorting. Indeed, each of the processes mentioned above delivers a variety of objects. At our scale, they look alike, but at the atomic level they are all different. Current solutions involve an entire toolbox of physical and chemical cues that play on their selective affinities in order to sort them out.

In short, it took about 20 years of research to learn how to work with this material: how to handle it, how to deal with its environment and how this environment acts in return on its properties, and finally, how to mass-produce components made of a large number of CNTs. It is after all a relatively short period of time compared to that required for the development of electronic silicon materials (it took no less than 40 years to master the synthesis of very large ultra-pure silicon crystals).

However, faster mastery doesn't mean equal mastery. Even with the improvements in the industrial production of carbon nanotubes, these materials are not about to overthrow silicon. They are rather meant to complement it.

6. A Disruptive Innovation: Optoelectronics

The highest promises of CNTs rest on their special and unique performances in optoelectronics. Semiconductor CNTs interact strongly with light. They absorb light of certain colors, in specific wavelengths such as near infrared, which is of particular interest for medical imaging (living tissues are relatively transparent in the near infrared) or, again, for photovoltaic energy.

CNTs can also emit light in specific ways that depend on their geometry. In this respect, these materials are paradigmatic of nanoscience in the sense that their properties and performances depend on their interaction with the environment as much as on their chemical composition. Depending on which atoms are (intentionally or not) attached to the surface of the object, its behavior will differ. If we do not take care, it will interact randomly with the environment and this interaction will modify its properties. Smalley's team learned how to take care of these tiny objects, for example, by encapsulating them into a small protection bubble, a micelle, which isolates them from the rest of the world, so that they agree to show their optical properties. These studies demonstrated that their optical behaviors are far more diverse than those predicted by theory because of coupling effects with their milieus [12].

The promising optoelectronics applications of CNTs are the so-called "hybrid" photonics, that is, integration of CNTs *on silicon*. As for graphene, which has been touted so much over this last decade as the future material for the post-silicon age, it has no band gap and is thus not a semiconductor; it is therefore not further destined to replace silicon, but rather to complement it with high-frequency applications. Noticeably, nanotube-chip designers [9] never compared the performance of the carbon chips against equivalent silicon chips, which benefits from more than 50 years of research and development fueled by huge commercial returns. They rather insist that carbon chips can be built using existing technology and integrate with the mainstream of silicon. For not only is it difficult to radically change a technology that has taken so long to develop, it is also crucial for a new technology to connect with the old and integrate a diverse ecology of materials. The realization of all-carbon computers would require an alternative industrial ecosystem [13].

Finally, nanocarbons are less to be seen as competitors — between themselves as well as between them and silicon — than as mediators or go-betweens. They pave the way to an entire class of low-dimensional materials other than carbon: in the one-dimension, to nanowires, and other nanotubes, such as manganese vanadium oxide nanotubes; in 2D, to giant flat molecules: boron nitride (graphyne), silicene, borophene, germanene, phosphorene, etc. Nanocarbons make the nanoworld proliferate because the material ways of being they have initiated are to some extent transferable to other elements. Even silicon can form nanotubes!

7. Health Promises and Risks

Because carbon is the main element of biological life, fullerenes and carbon nanotubes have first been naturally considered biocompatible. Moreover, due to carbon's genius of association, nanocarbons can be quite easily functionalized. They are grafted with other compounds at their surface such as therapeutic agents, pH-sensitive molecules, DNA or peptide. They accordingly qualify as good candidates for drug, gene or protein delivery in the cell. Targeted drug delivery has become a popular concept in nanomedicine and is undoubtedly the most successful and uncontroverted application of nanotechnology. Being extremely stable physically and chemically, CNTs are not easily biodegradable.[4] This is seen as an asset for traveling unscathed a path strewn with pitfalls through the bloodstream and various intercellular milieus. Their mechanisms of internalization in the cell vary according to geometry, size, charge density and

---

[4]Except for some bacteria and fungi. However, some studies show that there are some mechanisms of CNT clearance by macrophages in the rat lung (but not yet, so far, in the liver). The CNTs are chemically modified and cleaved, and then metabolized into innocuous carbohydrates.

functionalizing agents. CNTs can cross membranes by passive diffusion, acting as nano-needles, or by energy-dependent mechanisms involving membrane deformation processes: the cell membrane can then form vesicles, small bubbles which encapsulate the nanotubes and travel inside the cell before releasing them with their therapeutic agents.

In addition, thanks to their capacity of light-absorption in the near-infrared, they can be used to damage diseased cells by heating. This photothermal therapy can be combined with their use as imagery agents, providing precise probes to enlighten specific zones of the body's inner space (theragnostics, i.e. therapy + dia/prognostics). Their potential in nanomedicine derives once again from their conjunction of original properties and their ability to infiltrate various milieus (here living tissues).

However, the same wonder virtues, in particular their capacity to cross biological barriers, raise serious toxicological concerns. In particular, some studies have shown structural and toxicological similarities between CNTs and asbestos fibers [14]. Yet, in order to study their toxicity it is not enough to consider the chemical nature of the material (carbon, biocompatible). Nanotubes and nanomaterials in general induced a paradigm shift in toxicology: from the dogma 'the dose makes the poison' to a much more complex view of sources of toxicity that raise endless controversies. Because of their size first nanotubes have been proved cytotoxic by some studies, while others have concluded to a high pulmonary toxicity independent of length [15, 16]. But also: the type of exposure (ingestion, inhalation, epidermal contact), the detail of the crystalline structure, the geometrical shape, their relations between them (their agglomeration behavior, for example), with the co-agents (e.g. the catalysts required for the synthesis of CNTs sometimes remain as residues in the final product) and with the environment, and the way this relation can be tuned by proper functionalization. For instance, nanotubes tend to agglomerate in aqueous environment but can be rendered soluble by coating them with micelles. Each variation of these factors completely changes their toxicological effect. It is thus not the *substance* of the object (its intrinsic properties) that matters but its *reliance*. It is the coupling of certain specific features of the object with some of its milieu of operation that redefines the nano-object as an object of reliance, and, I would say, as an object of care.

As a result, there is no simple answer to the question of toxicity. Nanotoxicological studies look on a case-by-case basis at each type of nanomaterial in its specific milieu of operation and use. In some cases, however, good cooperative relationships have been established between nanocarbons and living cells, for instance, in restoring damaged nervous tissues, which is of interest for spinal chord or brain regeneration after injury [17]. The nanotubes play two roles here, the first relates to their morphology, the other to their electronic properties. On the one hand, it turns out that they constitute a favorable substrate for the growth of neurons in vitro. They "feel good" on contact with the nanotubes and reweave their damaged fibrous network along them. On the other hand, the electronic properties of nanotubes help distant neurons to reconnect with each other and recover electrical function even when the biological nerve itself is still damaged — another illustration of there reliance. These two mutually supportive processes are strongly dependant on the type of nanotube (single or multi-walled, metallic or semiconductor, etc.). As a result, the potential developments of carbon nanostructures for neurobiology are now rapidly expanding from molecular imaging to neuro-regenerative scaffolds and graphene-based neural interfaces.

However, in vivo uses of nanocarbons require special safety measures and clinical trials. They have to be highly purified, strictly characterized and standardized. Yet, toxicological studies of CNTs often remain controversial because CNTs interact with the biological milieu, which is never fully controllable. Therefore, despite great advances nanocarbons are still essentially promising materials.

To be sure, many of their potential applications are likely to remain speculative, doomed to feed the kitsch retro-futurism of tomorrow. But these wonder materials and their ways of being "materials" are no less intriguing. For it is also and perhaps foremost the different kinds of carbons' worldliness at this scale that justifies the phrase "nanoworld", as a world of endless production of allotropes.

References

1. Bensaude-Vincent, B., Loeve, S. (2018). *Carbone. Ses vies, ses œuvres.* Paris: Le Seuil.

2. Kroto, H. W., Heath, J. R., O'Brien, S. C., Curl, R. F. and Smalley, R. E. (1985). C$_{60}$: Buckminsterfullerene. *Nature*. 318(6042), 162–163.

3. Smalley, R. E. (1997). Discovering the fullerenes (Nobel lecture). *Angewandte Chemie International*. 36(15), 1594–1601.

4. Edwards, B. C. (2000). Design and deployment of a space elevator. *Acta Astronautica*. 47(10), 735–744.

5. Mody, C. M. (2008). 1990: Buckyball & Carbon nanotubes. In Giguere, R. J. (Ed.) *Molecules that Matter*. Philadelphia: Chemical Heritage Foundation. pp. 159–176.

6. Monthioux, M. and Kuznetsov, V. (2006). Who should be given the credit for the discovery of carbon nanotubes? *Carbon*. 44(9), 1621–1625.

7. Endo, M. (2002). Interview by Bernadette Bensaude-Vincent [Internet]. *Sciences*. Histoire orale. https://www.sho.espci.fr/spip.php?article48.

8. Iijima, S. (1991). Helical microtubules of graphitic carbon. *Nature*. 354(6348), 56–58.

9. Hills, G., Lau, C., Wright, A., Fuller, S., Bishop, M. D., Murphy, D., *et al.* (2019). Modern microprocessor built from complementary carbon nanotube transistors. *Nature*. 572(7771), 595–602.

10. Novoselov, K. S., Geim, A., Morozov, S. V., Jiang, D., Zhang, Y., Dubonos, S. V. *et al.* (2004). Electric field effect in atomically thin carbon films. *Science*. 306(5696), 666–669.

11. Ricaud, M., Lafon, D. and Roos, F. (2008). Les nanotubes de carbone: Quels risques, quelle prévention? *INRS Hygiène et sécurité du travail*. 206(8), 43–57.

12. Bachilo, S. M., Strano, M. S., Kittrell, C., Hauge, R. H., Smalley, R. E. and Weisman, R. B. (2002). Structure-assigned optical spectra of single-walled carbon nanotubes. *Science*. 298(5602), 2361–2366.

13. Mody, C. M. The diverse ecology of electronic materials. In Teissier, P., Mody, C. M., van Tiggelen, B. (Eds.) *From Bench to Brand and Back: The Co-Shaping of Materials and Chemists in the 20th Century; Cahiers François Viète*. (2017). III(2), 217–241.

14. Poland, C. A., Duffin, R., Kinloch, I., Maynard, A., Wallace, W. A., Seaton, A., Donaldson, K., et al. (2008). Carbon nanotubes introduced into the abdominal cavity of mice show asbestos-like pathogenicity in a pilot study. *Nature Nanotechnology*. 3(7), 423–428.

15. Kostarelos, K. (2008). The long and short of carbon nanotube toxicity. *Nature Biotechnology*. 26(7), 774–776.

16. Simon-Deckers, A., Gouget, B., Mayne-L'Hermite, M., Herlin-Boime, N., Reynaud, C., Carrière, M. (2008). In vitro investigation of oxide nanoparticle

and carbon nanotube toxicity and intracellular accumulation in A549 human pneumocytes. *Toxicology.* 253(1–3), 137–146.

17. Fabbro, A., Villari, A., Laishram, J., Scaini, D., Toma, F. M., Ballerini, L., *et al.* (2012). Spinal cord explants use carbon nanotube interfaces to enhance neurite outgrowth and to fortify synaptic inputs. *ACS Nano.* 6(3), 2041–2055.

Part 3

Controversial Materials

Bernadette Bensaude-Vincent

The life stories gathered in this section point to the ambivalence of materials. The more interesting and attractive a material is, the more it is prone to raise controversies. For instance, chlorine and asbestos are both common, cheap and afford high performances. Chlorine distinguishes itself by its activity and reactivity at high concentrations, white lead by its strong coloring power, asbestos by its unique combination of insulation and flame resistance properties, plutonium by its capacity of releasing high energy after fission of its atomic nuclei. However, these "wonder materials" raise serious concerns because of their health and environmental impacts. They exemplify the need for an extension of materials thinking beyond the traditional system focused on performances and process. They call for a more holistic approach taking into account the full life cycle of materials, their impacts on users and on the environment. Hazards, risks and uncertainties are integral parts of materials thinking.

This section raises an intriguing issue about the relation between the production of knowledge and technological choices or decisions. Even though technology cannot be considered as applied science, in most industrial societies, technological choices are supposed to be rational. They tend to be based on sound evidence and assessed through a balance between costs and benefits.

Yet, all the examples in this section suggest that the production of scientific evidence about health or environmental damages is not necessarily followed by reasonable decisions. Although the toxicity of white lead had been scientifically established and the health damages of asbestos on miners and workers publicized for decades, these materials were not banned and remained widely used in construction. Plutonium is special since it has been specifically made to produce destruction in war contexts. Knowledge about its toxicity and effects on workers' health has been deliberately kept secret for many years through a strict censorship. But secrecy is not the unique obstacle to prevent the damages due to toxic materials.

Apart from crude censorship, public scientific evidence about the toxicity of a material is not always effectively followed by appropriate measures of protection. Epidemiological studies often generate a scientific controversy about the evidence of risks and hazards. Such debates point to the complex dynamics involved in the construction of scientific evidence. Indeed, scientific controversies are integral part of the regime of normal science. For instance, the controversy surrounding the "birth" of chlorine as a simple substance is presented by Hasok Chang as a standard example of how theoretical choices are informed by values and underdetermined by experimental evidence. But the life story of organic chlorine compounds such as DDT suggests that scientists can be encouraged to produce counter-evidence by industrial and political interests. The attempts at generating doubts through controversy maneuvers are designed for extending the social life of hazardous substances such as tobacco or bisphenols [1].

Even more puzzling than the generation of doubts and organized skepticism for destabilizing the evidence of harms is the fact that there is a more radical strategy of deliberate ignorance [2]. The stories of white lead and asbestos suggest that scientific evidence has sometimes been deliberately neglected, discarded or concealed. Emmanuel Henry argues that there is no correlation between the production of scientific evidence of the impact of asbestos on occupational health and its declining use in the recent decades.

The time lag between the production of scientific evidence about the toxicity of controversial materials and the political measures to prevent environmental damages or protect the public health provides a good probe into the value priorities that rule a society. For example, lead white, which continued to be widely used for more than a century in the manufacture of paints even though its harmful effects were known and established, is a clear indicator that economic profit takes precedence over public health. As long as the profits of chemical lobbies are more important than the health of bees, workers and farmers, toxic materials will have a long life.

By displaying conflicts of values, this section will alert future materials scientists and engineers about their social responsibility in a complex world where their technical expertise may be instrumentalized by corporation lobbies and political agendas capable of distorting risk assessment and blocking regulations.

References

1. Oreskes, N. and Conway, E. (2010). *Merchants of Doubt. How a Handful of Scientists Obscured the Truth on Issues from Tobacco Smoke to Global Warming.* New York: Bloomsbury.
2. Proctor, R. and Schiebinger, L. (Eds.) (2008). *Agnotology: The Making and Unmaking of Ignorance.* Palo Alto: Stanford University Press; Boudia, S. and Jas, N. (Eds.) (2014). *Powerless Science? Science and Politics in a Toxic World.* New York, Oxford: Berghahn Books. p. 280.

Chapter 9

Chlorine

Hasok Chang

University of Cambridge, UK

1. Introduction

If chemical elements were people, chlorine would be a celebrity. With the combination of its high reactivity and its relative abundance on Earth, chlorine has been making headlines throughout its life. Chlorine gas still has emblematic status as the first major chemical weapon ever deployed, in 1915. On the other hand, chlorine and various chlorine compounds have also worked as life-saving disinfectants; the familiar and distinctive smell of chlorine in swimming pools and household bleaches is a reminder of their widespread practical use. The political controversy surrounding chlorine continued long beyond the First World War, particularly in relation to the use of organic chlorine compounds, ranging from the herbicides used by the U.S. forces in Vietnam (including the infamous Agent Orange) to DDT and other insecticides whose environmental impact was hotly debated following the publication of Rachel Carson's *Silent Spring* in 1962.

Chlorine has been surrounded by controversy right from the moment of its discovery. Carl Wilhelm Scheele, who first isolated what we now call chlorine in 1774, did not think of it as an element, and called it "dephlogisticated muriatic (marine) acid". When Antoine-Laurent Lavoisier and his colleagues overturned the phlogiston theory, they re-named Scheele's substance "oxygenated muriatic acid". "Marine acid" and "muriatic acid" were both common longstanding terms designating what we now call hydrochloric acid, due to its original manufacture by alchemists by reacting sea salt with vitriolic (sulfuric) acid. Consensus on the elementary nature of chlorine only came with Humphry Davy's work in the 1810s. Afterward chlorine also featured prominently in other debates in theories of matter, including Prout's hypothesis and the solar neutrino problem.

Such a controversial material substance creates a wonderful opportunity for an object biography. In this chapter, I offer some reflections arising from a collective undergraduate research project that I directed at University College London, which resulted in a biography of chlorine [1]. I will share some thoughts about what made chlorine a good subject for an object biography, and illustrate the points with some of the stories from our work. I will close with some reflections on the pedagogical experiment through which our work was carried out.

2. The Cogency of a Material Biography

Following the life of such a controversial material substance as chlorine creates a wonderful conceptual space in which to explore the development of science, technology, medicine in their full social, intellectual and material contexts. Before engaging in the project on chlorine, I had been inspired by some very informative and enjoyable works of object biography, the first of which was *Cod* by Mark Kurlansky. But even more intriguing to me were titles that traced the lives of basic material substances, including phosphorus, water, salt, hydrogen and the electron [2].

But are object biographies merely amusing? Is "biography" just an attention-grabbing metaphor for designating the history of an object? Taking the notion of biography literally at least for a moment, we need to ask whether a material object or substance can in any serious sense be taken as a protagonist, and whether it can be considered to have a "life". We should also be asking why it is that we write and read biographies of people, anyway — in other words, why should a person-centered narrative be considered an important genre in historiography? Would it not be better to keep the focus on events, trends, groups and institutions? Doesn't history told through the life of an individual present an overly idiosyncratic perspective? My sense is that it is exactly the idiosyncrasy of perspective that we should value in biographies. In the story of an individual's life, we see different aspects and elements of social and material reality converging and interacting with one another. This is a perspective that we do not get in other types of histories. It could be argued that what ultimately matters in our understanding of history is how life was lived, and lived experience belongs to individuals, albeit individuals who are fully immersed and situated in their social and physical environments.

With those thoughts, let us return to the question of object biographies and our particular case of chlorine. Chlorine makes an interesting and instructive protagonist, because it has a distinctive *character*, which can be summed up as a high degree of reactivity. And this character has been all the more impactful because chlorine is actually quite abundant on earth (unlike, for example, fluorine or even phosphorus, which could be ignored for most purposes). For most of human history chlorine exerted its influence in a hidden way, by its constitution of the common salt (sodium chloride, $NaCl$) and other compounds. The high reactivity meant that chlorine would almost always cling tightly to other substances and could only be isolated through very particular operations. When chemists discovered those operations in the late 18th century, chlorine as a pure substance was born — one might ask whether pure chlorine gas exists stably in any appreciable quantity in nature at all, and at least I am not aware of a situation in which it does. Immediately upon its birth chlorine exhibited its reactive character: destroying colors, smells, germs and larger organisms. So began the multi-pronged career of chlorine in practical applications, as bleach, disinfectant and poison gas. Various compounds of chlorine also exhibited particular destructiveness, giving rise to their use as insecticides and herbicides.

Chlorine's reactivity also allowed it to have a distinguished career as an agent of chemical research, especially through the chlorine–hydrogen substitution reactions which helped so much in the understanding of valency, and the elucidation of the structure of organic molecules. Sadly, this was an important phase of chlorine's life that we did not manage to include in our biography. However, good historical accounts of that development exist elsewhere [3]. Interestingly, in this regard chlorine was so useful because it was actually not *quite* so reactive as some other elements, or rather, not so reactive in a physically destructive way, so it could be handled easily by reasonably well-trained researchers. An obvious contrast is with potassium and sodium, which did not make handy laboratory tools because they became oxidized so easily and burst into flames on contact with water.

So, we can see that the specific character of chlorine as a material shaped the story of its life in very particular ways, similarly to how the character of the protagonist in a personal biography shapes the events that form the narrative. But any attentive reader of our biography of chlorine will also notice that there are some stories in it that do not quite fit into the overarching theme of chlorine being an "element of controversy" due to its highly reactive character. Particularly, the role of chlorine being a focus of the dispute concerning Prout's hypothesis (that all atomic weights were whole-number multiples of hydrogen's, or even that all atoms were made up of hydrogen atoms) was due to its atomic weight being stubbornly just around 35.5, which was due to an accidental fact, namely that the chlorine we have on Earth is roughly a 3:1 mixture of two isotopes, chlorine-35 and chlorine-37, with those atomic weights. Likewise, chlorine was a crucial element in the story of neutrino detection and solar astrophysics because it just happens that one of its isotopes (chlorine-37) interacts relatively well with neutrinos and forms a stable and isolatable product (argon-38). This property of chlorine has nothing to do with its chemical reactivity. So, it could be argued that at least part of the tales collected in our biography of chlorine do not really fit in. But we must ask: does everything fit so tidily even in a personal biography? Surely all sorts of accidental circumstances impinge on a person's life, and a good biography should tell about those circumstances and their consequences, too. And it is also important to note how various accidents can serve to enhance a protagonist's distinctive character, or dull its edge. The circumstances just mentioned were accidents that happened to enhance the controversial tenor of the life of chlorine.

To illustrate in more detail the kind of stories that the perspective of material biography facilitates, I will now present three vignettes from our biography of chlorine. In presenting very brief versions of these fascinating and intricate stories, I am giving a perspective on what makes a material biography work, rather than focusing on the main features that the authors themselves foregrounded.

3. The Difficult Birth of Chlorine

The first story is about the "birth" of chlorine as a pure substance, investigated by Ruth Ashbee [4]. The high reactivity of chlorine not only made its isolation non-trivial but created great theoretical difficulties in the interpretation of the relevant experiments. A very important accidental factor in the story is that the first isolation of chlorine took place in the context of the upheaval in chemistry that we now call the "Chemical Revolution". The Swedish pharmacist and chemist Scheele, who incidentally also made oxygen long before Lavoisier, made chlorine by reacting muriatic acid (our hydrochloric acid, HCl) with "manganese" (mostly our manganese peroxide, MnO_2). Scheele designed and interpreted many of his pioneering experiments using the concept of phlogiston (similarly as Joseph Priestley in England), and understood this reaction as the removal of phlogiston from the acid by the manganese, calling the gaseous product of the reaction "dephlogisticated muriatic acid".

Lavoisier and his fellow "antiphlogistians" in France could not tolerate Scheele's phlogiston-based interpretation. For them muriatic acid did not contain phlogiston because phlogiston didn't exist, and what Scheele made from muriatic acid couldn't be dephlogisticated muriatic acid. An important general interpretive strategy they employed was the following: where phlogiston theorists claim that they see a removal of phlogiston, what is actually going on is the addition of oxygen. Accordingly, Claude-Louis Berthollet fitted Scheele's chlorine experiment into Lavoisier's framework by saying that chlorine was the product of the addition of oxygen (from Scheele's "manganese") to muriatic acid. So he re-named Scheele's "dephlogisticated muriatic acid" as "oxygenated muriatic acid", or "oxymuriatic acid".

But how could anyone tell which of these interpretations was correct? (Modern chemists will look on in horror: *both* Scheele's and Berthollet's interpretations are completely wrong!) Berthollet set about gathering other evidence to decide the verdict. One of the experiments he cited is quite intriguing: if a solution of chlorine gas in water is exposed to sunlight, oxygen gas and muriatic acid are produced, showing that oxy-muriatic acid (chlorine) is indeed a compound of oxygen and muriatic acid. This phenomenon is real, but we now say that what happens here is the dissociation of water, which releases oxygen and leaves hydrogen behind to combine with the chlorine to form hydrochloric acid. This is ironic, given that in other situations Berthollet and other Lavoisierians were quite fond of pointing out that the phlogiston theorists were fooled by the action of ever-present water. Early experiments involving chlorine definitely suffered from what philosophers of science call the "underdetermination of theory by evidence".

In the next stage of the history, the argument was between the Lavoisierian account and the new hypothesis that chlorine was an element. Berthollet, still defending the notion of chlorine as oxymuriatic acid, interpreted the oxidization of metals in the presence of chlorine as evidence that chlorine contained oxygen. But as ever, there was water in that experiment, too, and the advocates of elementary chlorine could easily make sense of it as the chlorine taking hydrogen from the water to form hydrochloric acid, and the oxygen from the water forming an oxide with the metal. How Davy convinced the chemical community about the elementary nature of chlorine is an intricate story, told in detail by Tamsin Gray, Rosemary Coates and Mårten Åkesson in our biography of chlorine [5].[1]

4. The Strange Case of "Chlorine Chambers"

Our next vignette is the surprising story of how chlorine gas was used as a therapeutic agent in the immediate aftermath of its debut as a dreaded agent of chemical warfare, told by David Nader and Spasoje Marčinko [6]. Chlorine gas was touted as a cure and preventative for respiratory diseases in the early 20th century in the United States, endorsed even by President Calvin Coolidge. Our research on this episode began with a few mysterious lines I had stumbled upon in a popular chemistry book: "Very small amounts of chlorine in the air help to ward off colds and to relieve them when once they have gained a hold" [7]. The author gave no sources, and the historians

---

[1]An important part of this story was the resulting demise of Lavoisier's theory of acids, which saw oxygen as the essence of acidity.

of medicine that I consulted had not heard of this episode. Nader and Marčinko unearthed the story of how chlorine therapy was promoted by the U.S. Chemical Warfare Service (CWS) to aid its own survival.

This case shows how complex the consequences of the basic character of a material can be, as they become refracted through powerful social circumstances. Our story begins in 1918–1919, when the First World War was coming to a close and the world was hit by the "Spanish flu" pandemic that resulted in 20 million deaths worldwide at a conservative estimate. Various medical officers during the war had made anecdotal observations that frontline soldiers and workers in factories manufacturing chlorine gas seemed to be less prone to catching the flu. These chance observations resonated with the longstanding knowledge that chlorine worked as a disinfectant. Could chlorine kill the flu germ in the environment, or even in the body? In the absence of any effective means of fighting the pandemic, even the American Chemical Society seriously debated the use of chlorine therapy at its 1919 meeting, where the favorite phrase of the day reportedly was: "Try it — it can do no harm." (The uncomfortable parallel with the current coronavirus pandemic is all too apparent.)

This situation was seized upon as an opportunity by the leadership of the U.S. CWS, especially its Director, General Amos Alfred Fries, who believed that chemical weapons were morally no more abhorrent than conventional weapons. The CWS was facing the threat of extinction in the postwar revulsion against chemical warfare that would soon lead to the establishment of the Geneva Protocol. Fries launched a campaign to advertise the beneficial uses of chemical-warfare agents, and in 1922 he appointed the well-respected military physician Edward Vedder to take charge of investigations into the medical uses of chlorine gas. The CWS treated a large number of patients suffering from various respiratory diseases, and reported high rates of success. Fries's campaign included the cooptation of the political elite, and he managed to get a chlorine chamber installed in the U.S. Capitol building, as a convenient and comfortable setting for the administering of low-concentration chlorine gas. Chlorine therapy became highly fashionable for a time, especially after President Coolidge's treatment became front-page news in May 1924; at least 169 members of Congress received chlorine treatment, and the public clamored for an affordable version.

A modern commentator would expect that there must have been a swift scientific refutation of the claims of chlorine therapy, and apparent evidence for that can be found. Experts in the New York City Department of Health and elsewhere launched strong critiques of the CWS studies, pointing out the lack of control groups and the careless use of statistics. Even Vedder himself seems to have lost the zeal in his promotion of chlorine therapy within a few years. Yet, it was not straightforward scientific critique that was chiefly responsible for the demise of chlorine chambers. It seems that the CWS gave up on the promotion of chlorine therapy so easily only because they found a better public-relations vehicle: insecticides. The latter story has been documented and analyzed extensively by Edmund Russell [8]. Meanwhile, General Fries also diversified his PR portfolio, extolling the peacetime use of chemical weapons: "every week . . . gas is used by Police Departments [in] saving lives of policemen and innocent bystanders and in overcoming criminals in barricaded buildings without a shot being fired or without danger to anyone."

Through this narrative of the rise and fall of an unlikely medical intervention from a century ago, we can have a vivid experience of the complex interactions of science, medicine, politics and war. The destructive power of chlorine made it a potent agent both in warfare and therapeutics at a time of global crisis, placing it right in the midst of a storm that was political, medical, scientific and ethical all at once.

5. Organochlorines and the Environment

Moving into the second half of the 20th century, the controversial life of chlorine continued through the threats that it posed to the environment. This time it was not so much chlorine as a pure substance but "organochlorines", namely chlorinated organic substances, that were at the center of attention. So, here the consequences of chlorine's character are exceptionally complex, this time refracted through its material combination with other chemical substances as well as through various ecological, social and ethical contexts. Joe Thornton has issued a comprehensive warning about the prevalence of organochlorines in modern life and their devastating environmental impact, and proposed a new policy perspective for their control and management [9]. In our biography of chlorine we focused on two concurrent early controversies concerning organochlorines, showing the struggle for authority on scientific matters.

Sam Raphael and George Kalpadakis [10] charted the course of disputes concerning the use of organochlorines in the extensive program of defoliation and crop-destruction undertaken by the U.S. military during the Vietnam War. The infamous defoliant Agent Orange as well as Agent White were mixtures of organochlorines. Starting with a statement issued by the Federation of American Scientists in March 1965, many civil (non-governmental) scientists launched strenuous critiques of the use of organo-chlorine defoliants, challenging the views of the government and the scientists working within the government. The initial arguments were made on moral grounds, not challenging the assumption that the chemical agents were not harmful to people or the environment. Between 1965 and 1970, however, this presumption of harmlessness was increasingly questioned and undermined by civil scientists contradicting pronouncements by the Department of Defense and other governmental agencies. The pressure from the civil scientists culminated in the 1970 fact-finding mission to Vietnam organized by the American Association for the Advancement of Science and led by the Harvard biologist Matthew S. Meselson. Rather than confronting this onslaught, the government quietly phased out the defoliation program, whose military effectiveness was increasingly being questioned in any case. It is notable that this change was not caused by the production of significant new knowledge, but by the shift of authority from the government to the civil scientists.

While the debates about defoliants were driven mostly by the political and moral outrage concerning the Vietnam War, a more environment-focused fight was brewing also on matters of U.S. domestic policy. This is the story, told by Kimm Groshong [11], of the reception of Rachel Carson's 1962 best-seller *Silent Spring*. Carson used the example of the deleterious effects of DDT (dichloro-diphenyl-trichloroethane) to argue that the public should demand more investigation into the long-term effects of pesticides. She was certainly not the first person to write about this subject, but hers was the first comprehensive statement of the problem in plain terms for a broad readership. Carson's work elicited furious reactions from the chemical industry and its allies, accusing her of inaccurate and biased presentation of data, and portraying her as a scientifically unqualified fanatic. However, the general public's reaction to her work was resoundingly positive, and her scientific claims stood scrutiny sufficiently well. While many of those who criticized *Silent Spring* were enraged by Carson's attempt to alert the public to the power wielded by the growing science–industry–government complex, the public and a significant portion of the scientific community were increasingly willing to question those in positions of political and scientific authority. The bifurcated and vehement response to *Silent Spring* was a natural consequence of the context into which it was published.

6. The Process of Research and Writing

In closing I wish to comment briefly on the pedagogical experiment through which our biography of chlorine was produced, which was conducted at University College London from 2000 to 2005. In my teaching I always encouraged undergraduate students to carry out original research in my courses, and even unexceptional students often rose to the challenge in some small ways. However, I also came up against a frustrating problem: after assessment, these valuable essays simply sat in piles collecting dust. The reality was that most students within the confines of a course-unit or even a final-year dissertation were not quite able to bring their research to the publishable level. This frustration served as a catalyst for an innovation, which I called the mechanism of "inheritance": each cohort of students passed on to the next cohort the fruits of their work for further improvement. This was continued for five years, until enough publishable work was produced.

Not all cases of inheritance worked out in the linear fashion that I had envisaged. Sometimes a single student managed to finish up a topic to a very high standard. Often more than one student wanted to work on a given topic in the same year, requiring a division of labor in addition to a sequential cumulation. The mode of work and practical arrangements had to be crafted with some creativity. For example, we were stuck with having to give an exam in the course, but we turned this annoyance to our advantage by using the exam as a way of forcing students to learn about each other's projects; this helped us a great deal in turning the group into a real community. Also helpful for community-building were small-group tutorials that I ran with students working on similar topics. And the inheritance mechanism was strengthened greatly by handing down all the reading materials and research notes as well as the final product from each student, and to help students take that seriously I gave marks on the research portfolio as well as the final essay [12].

The whole process confirmed and strengthened my belief that learning can take the form of knowledge-*production*, going beyond knowledge-*acquisition*. Despite many recent efforts at reform, our educational systems still tend to be based on the notion that students are there to be trained by passively acquiring pre-existing knowledge, before they can produce original work. In contrast, I believe that the processes of learning and original research can be soldered into one at an early stage. Similarly, even pioneering programs of undergraduate research tend to make it a preserve of the best and the brightest. In our project we made original research a routine part of ordinary undergraduate work, from which any competent and hard-working student could benefit.

To make our project viable, a flexible and multi-faceted theme was required. Attracting enough students to the project required accepting students with a broad range of backgrounds and interests, and creating a coherent community out of them. The genre of object biography was very effective for this purpose. Even where the biographical metaphor or analogy does not quite hold, the object-focus created an ideal interdisciplinary forum, bringing together historical, philosophical, sociological and political perspectives, and building an easy bridge between the humanities and the STEM subjects.

References

1. Chang, H. and Jackson, C. (Eds.) (2007). *An Element of Controversy: The Life of Chlorine in Science, Technology, Medicine and War* London: British Society for the History of Science.
2. Kurlansky, M. (1997). *Cod: A Biography of the Fish that Changed the World.* New York: Penguin; Emsley, J. (2000). *The Shocking History of Phosphorus: A Biography of the Devil's Element.* London: Macmillan; Ball, P. (1999). *H₂O: A Biography of Water.* London: Weidenfeld & Nicolson; Multhauf, R.

(1978). *Neptune's Gift: A History of Common Salt.* Baltimore: Johns Hopkins University Press; Rigden, J. S. (2002). *Hydrogen: The Essential Element.* Cambridge, Mass.: Harvard University Press; Arabatzis, T. (2005). *Representing Electrons: A Biographical Approach to Theoretical Entities.* Chicago: University of Chicago Press; Daston, L. (Ed.) (2000). *Biographies of Scientific Objects,* Chicago: University of Chicago Press.

3. Rocke, A. J. (2010). *Image & Reality: Kekulé, Kopp and the Scientific Imagination.* Chicago: University of Chicago Press, p. 9ff.

4. Ashbee, R. The discovery of chlorine: A window on the chemical revolution. In Chang, H. and Jackson, C. *Element of Controversy* (note 1), pp. 15–40.

5. Gray, T., Coates, R. and Åkesson, M. The elementary nature of chlorine. In Chang, H. and Jackson, C. *Element of Controversy* (note 1), pp. 41–72.

6. Nader, D. and Marčinko, S. The rise and fall of 'chlorine chambers' against cold and flu. In Chang, H. and Jackson, C. *Element of Controversy* (note 1), pp. 296–323.

7. Newton Friend, J. (1961). *Man and the Chemical Elements.* 2nd Ed. London: Griffin, p. 48.

8. Russell, E. (2001). *War and Nature: Fighting Humans and Insects with Chemicals from World War I to Silent Spring.* Cambridge: Cambridge University Press.

9. Thornton, J. (2000). *Pandora's Poison: Chlorine, Health and a New Environmental Strategy.* Cambridge, Mass.: The MIT Press.

10. Raphael, S., Kalpadakis, G. and O'Reilly-Weinstock, D. War and the scientific community. In Chang, H. and Jackson, C. *Element of Controversy* (note 1), pp. 333–353.

11. Groshong, K. The noisy reception of silent spring. In Chang, H. and Jackson, C. *Element of Controversy* (note 1), pp. 360–382.

12. For further details on the pedagogical process, see Chang, H. and Jackson, C. *Element of Controversy* (note 1), pp. 383–394; Chang, H. (2009). Chlorine: Undergraduate research on an element of controversy. *Journal of Chemical Education.* 86, 418–420.

Chapter 10

White Lead

Judith Rainhorn

Université Paris 1 Panthéon-Sorbonne
Centre d'histoire sociale des mondes contemporains
(UMR 8058), Maison française d'Oxford

1. Introduction

White lead is an appropriate material for a biography. This millennia-old substance was known to the Persians and the Ancient Greeks. The method of producing it did not change considerably over time; from the recipe described by Pliny the Elder (1st century BCE) to the process used by Northern European industrialists in the early 19th century, what changed was mostly the scale of production. The industrial manufacturing process — known as the "Dutch process" as it was developed in workshops of Amsterdam and Rotterdam in the late 18th century — mainly consisting of economies of scale in enlarged workshops, allowed a significant increase in production of one of the luxury materials that had ensured Venetian trade dominance in the late Middle Ages [1]. In the 19th century, lead oxide was mainly produced as a pigment for building paint — as the building sector grew along with the widescale urbanization of Europe — and for many other industrial sectors (porcelain, wallpapers, glossy paper, artificial flowers, etc.).

This biography is a portrait of white lead as a serial killer, and a story of the attempts to halt its deadly effects [2]. It stretches across the millennia, over alternating highs and lows, with brief periods of fame and longer periods in the shadows. After a lengthy "childhood" during which actors and prostitutes used it to paint their faces, later becoming a celebrated element of pharmacopeia for whitening birthmarks and soothing the painful nipples of nursing women, white lead finally came of age in the early 19th century — before disappearing from the scene mid-20th century. Considering this long-term economic cycle, two contradictory biographies of white lead can be written, a light narrative and a dark one: manufactured on an industrial scale, it was attributed all the virtues of a paint pigment that enjoyed booming commercial success thanks to its intrinsic properties (whiteness, opacity, resistance, ease of drying and low cost); but it was also criticized as a poison responsible for the sickness and death of thousands of people. In work spaces, white lead is responsible for lead poisoning, a slow intoxication that affects the nervous system, brain and kidneys, causing serious and recurring attacks that can lead to death. In the archives of chemical industrialists, world fair brochures, and paint manufacturers' accounts, full of orders, we can trace its bright face. Meanwhile, the dark legend loomed in hospital admissions registers, the records of mutual insurance societies that compensated sick workers, physicians' accounts denouncing the ravages of lead poisoning in the working classes, or the workers themselves describing the "slaughterhouses" where they manufactured the deadly substance.

The invention of the "Dutch process" marked a major turning point: rolled sheets of lead were oxidized by contact with vinegar, heated steadily by horse manure. The process created a very fine white powder, which, mixed with oil, became the white paint that would soon be commonly used for the inner and outer walls of buildings and homes across Europe, as well as carriages and ships. White lead production became significant in Holland in the 1770–1780s, then in England and Belgium in the 1800s, in France in the early 1820s and in the U.S. just later [3]. France had a special role in this global picture as this material attracted so much scientific, political and public attention initially that France soon became the largest consumer of white lead in Europe. Therefore, the French case is both exceptional and emblematic, as most other countries that used it did not make it a *political* product until its disappearance from the market in the 1950–1960s.

Two major events could have hampered this massive growth of European and global white lead markets during the 19th century. First, its harmful effects upon the workers who manufactured it or the painters who used it: these effects were known in the 18th century and widely documented from the 1830s on, just after its economic growth cycle began. Second, an alternative substance, completely innocuous for the workers who manufactured or used it — zinc oxide — was invented in the late 18th century and already presented as a useful substitute to avoid the hazards of white lead. It nevertheless failed to overtake the chemical pigment market. The two substances would come into fierce competition in the 19th century and were eventually supplanted by new pigments after World War II.

Neither of these events undermined the commercial domination of white lead throughout the long 19th century (1780–1920). Why were health concerns pushed aside by industrialists, by the vast majority of paint manufacturers and users and by consumers among the general public? More specifically, how did technical innovation interact with risk acceptance and the rules governing the market and competition? The history of white lead forces us to set aside a teleological explanation whereby the social body slowly but inexorably evolved to take into consideration health and safety concerns. Instead, it shows the complex mechanisms at work in this conflict that intermingled scientific and technical knowledge, on the one hand, and price and economic interests, on the other hand. The history of this conflict, mixing public-health debates and market rivalry, is a perfect prism for analyzing industrial risk. It was not until the early decades of the 20th century — when the French parliament passed restrictive legislation in 1909, and the International Labour Organization implemented an international convention in 1921 — that white lead finally lost to its zinc rival. Nevertheless, it is not easy to disentangle which mechanism — legal bans, economic competition or technical advances — prompted the defeat of this long-life material. I argue that technical factors and the invention of new and improved pigments played as important a role as the development of restrictive legislation against the toxic material.

2. Entangled Whites: The Messy Chronology of a Scientific Controversy

When it entered into competition with white zinc, white lead raised a long controversy that threatened its lifetime as a commercial product. Let us survey this troubled period of its biography.

The first phase occurred in late 18th century. Paduan physician Bernardino Ramazzini had briefly described the use of white lead as a pathogen for studio painters, and French chemist Antoine-François Fourcroy, translating Ramazzini's opus in 1777, blamed the material in a lengthy note on the etiology of lead poisoning [4]. During this period, the famous chemist and encyclopedist Guyton de Morveau, assisted by pharmacist Jean-Baptiste Courtois, worked on creating a zinc-based white pigment that would be harmless and could replace white lead. Guyton de Morveau advocated substituting zinc for lead for both health and technical reasons, as lead paint tended to darken when exposed to sulfur vapors, whereas zinc paint appeared unaffected. Their research led to the first large-scale production of zinc white in France, while English chemist John Atkinson claimed this promising innovation as his own and patented zinc oxide in paint as a substitute for lead oxide (1796). Thus, at the end of the 18th century, in the two nations that used the most white lead — due to their early urbanization, as well as their sizeable naval and merchant-marine fleets which used white lead on a massive scale to paint ships — scientists were aware of the scientifically-proven toxicity of the material, and the existence of a safe alternative. Such eminent and renowned scientists as Fourcroy, Berthollet and Vauquelin all enthusiastically supported zinc white, the defects of which "are so slight compared to the disadvantages of using white lead, that its adoption cannot be reasonably refused" [5].[1]

However, the relatively high price of zinc white — three times more expensive than white lead at the time — overshadowed the health arguments and prevented the chemical innovation from becoming an industrial success. Conversely, white lead was widely promoted by industrial concerns and by the highest public health bodies in France (Conseil d'hygiène et de salubrité, as well as the prefectures and municipalities, which were responsible for enforcing health rules), in order to foster domestic production over imports from Holland, Belgium and England that weighed on the national budget. As a result, white lead, albeit acknowledged as a deadly

---

[1] Report by Messrs. Fourcroy and Vauquelin to the National Insitute (1808), cited in Jouffroy d'Abbans (1860).

poison by experts, kept its dominant position in the paint pigment market during the first half of the 19th century.

In the mid-19th century, another French innovation rekindled the economic rivalry and the scientific controversy. After several years of experiments, Jean-Edme Leclaire (1801–1872), a former painter who owned a paint company and was influenced by Saint-Simonianism, perfected a manufacturing process for zinc white that would allow for industrial-scale production and filed for a patent in 1845. Leclaire had witnessed the workers in Paris write the word "slaughterhouse" on the white lead workshop doors, and had at first stood by powerless to prevent the health effects of the "white poison" on his own workers. But from 1849, Leclaire's factory on the River Seine, a few miles downstream from Paris, was able to produce 6,000 kilograms of zinc white a day. Leclaire's zinc white was of higher quality than previous versions, more weather-resistant and its lower production cost made it a contender for industrial-scale use. Leclaire soon formed a partnership with the Société des Mines et Fonderies de la Vieille-Montagne [6],[2] the world leader in zinc mining based in Liege, Belgium, which had a powerful distribution network. He brought an industrial innovation to European markets that was a true competitor to white lead producers:

"It was a revolution in the habits of the trade; people were divided into two sides: those who remained attached to their old routine and used the poison, with gaunt, pale and sickly workers; and on the other side, the advocates of zinc white, whose personnel were plump, hearty men, with rosy cheeks and a winning appearance" [7].

Of course, this description, appearing in an etching given to Jean Leclaire by his grateful workers in 1851, was biased (Figure 1). The competition between the two products threatened the undisputed domination of white lead. The mid-19th century saw an increase in publications by reputed hygienists and professionals, such as Alphonse Chevallier (a member of the Paris Hygiene Committee) and the famous architect Viollet-le-Duc, in favor of zinc white [8]. In 1849, Jean Leclaire was awarded a gold medal by the Société d'Encouragement à l'Industrie nationale, and the following year, he received the Prix Montyon from the Academy of Sciences, followed by the Legion of Honor, the highest civil

---

[2] Subsequently called "Vieille-Montagne". See Peters (2016).

Figure 1. "Lithography given to Jean Leclaire by his workers, 1851" in Abel Craissac, *Conférence sur l'Empoisonnement des Ouvriers peintres par le blanc de céruse, donnée à la Salle du Crique de Saint-Quentin (Aisne) le 19 août 1905*, Puteaux, La Cootypographie. Public domain.

distinction in France. Ambitious public action coincided with this series of awards. In August 1849, while hygienist periodicals agreed that the manufacture and use of zinc white were harmless, a Decree, soon followed by a government circular in February 1852, called for zinc white to be used for all public building works.

At mid-century, all events suggested that white lead would soon disappear. Yet, a cloak of silence fell again on the dangers of white lead, which, like the many-headed Hydra, regained its strength even though it had been beheaded. I argue that the economic sector of the white lead chemical industry invented a new narrative to resist competition during this period and succeeded to make the issue dormant for nearly 50 years, before the issue of industrial poisons reentered the political debate at the turn of the 20th century.

Just after the 1898 Act establishing employers' liability for work-related accidents, the workers' movement to obtain compensation for occupational illnesses focused its efforts on industrial poisons. This chapter of the biography did not revolve around technical innovations, but was driven much more by changes in the labor movement, which enjoyed unprecedented ties with the political authorities under the French Republic. From 1900 to 1909, when the first law banning white lead paint was passed after years of parliamentary meanderings, white lead's public notoriety was at a peak, becoming a symbol. Labor leaders and journalists, high-ranking politicians such as Viviani and Clemenceau, public health experts and members of parliament worked together to make the perennial issue of industrial poisoning a major political "cause." In the French law passed just after World War I (Act of 25 October 1919), lead poisoning in painters was the occupational disease *par excellence* and its causes and effects were perfectly known by the general public, as newspapers were full of the health risk of using it and of economic rivalry of the two whites.

Back in the spotlight in 1921 in a transnational context, the conflict between lead-based and zinc-based paint briefly flared up again, as the 3rd International Labour Conference in Geneva debated an international convention banning white lead in building paint. France was clearly a pioneer on the topic, as the first country to have (partially) banned lead paint in buildings in 1909, and it was a leader in the ILO — whose first director was socialist Albert Thomas. The vigorous debates on white lead in Geneva read like a retrospective of the previous century: all the medical, technical, economic, health and humanist arguments were repeated in a single time and place. This repetition betrayed a collective amnesia making it necessary to review all the data already debated over the prior century: white lead was not dead yet. The 13th International Convention banning lead paint (1921), ratified by France, Belgium and a dozen other countries by 1926, gave a final twist to this debate, showing the powerful "manufacture of doubt" when assessing whether materials in the workplace are hazardous or innocuous.

3. Between Invisibility and Compartmentalization

For a century, the risk of industrial poisoning attracted sporadic attention from scientists, political authorities, experts and public opinion. Sometimes carefully documented, proven, debated and partially dealt with, the risk was in other times overshadowed or denied. The agnotological approach has been enhanced in recent social science research in the field of public health, and has shown that the invisibility of a public health issue is due not necessarily to deliberate concealment efforts, but most often to compartmentalization that prevents information from circulating outside limited, or even marginal, social spaces. The latter is a valuable perspective for casting light on the economic and scientific battle between white lead and zinc white, as periods of bright light and deep knowledge, on the one hand, and periods of limited circulation of the information, on the other hand, alternate during a century [9].[3] Thus, rather than speaking of periods of silence, one can identify periods of compartmentalization. Each time the white lead controversy returned to center stage during the 19th century, stakeholders' apparent amnesia regarding the material was quite surprising, as if the scientific and technical knowledge from previous episodes had been forgotten, as if new experts, chemists and physicians had to be brought onboard to prove the dangers of the material.

The first "silent period" on the issue of lead poisoning (1800–1845) in France was definitely part of the "erasure of the worker's body," a historical process identified by Le Roux in the dismantling of worker protections during the revolutionary period, giving rise to an industrialist mindset [10]. Prior to 1820, however, the economic stake was weak: production in France was still on a small scale and most white lead was imported. As a result, a small number of physicians and chemists regretted a health issue affecting a small number of people. The public authorities gave their full support to the first attempt to produce white lead on an

---

[3]The seminal work is Proctor and Schiebinger (2008). Also see Proctor (2012); Boudia and Jas (2013).

industrial scale and looked favorably on the factory built in 1809 in Clichy, on the outskirts of Paris. Given the French Empire's enthusiasm for industrial and chemical innovation, the white lead factory in Clichy fueled France's hopes of asserting its economic independence from rival countries. The support given to this enterprise from the outset was indicative of the overt collusion between politicians, industrialists and eminent public health experts, such as Chaptal, Thénard and Roard (the director of the Clichy white lead factory) [11]. Despite the development of white lead manufacturing in the 1820s and the subsequent increase in lead poisoning cases, which became apparent in the corridors of hospitals, the message of the material's dangers was restricted to the narrow spheres of public health. These health hazards were documented in the *Annales d'hygiène publique et de médecine légale* but found few echoes outside this specialist publication. Despite a few isolated cases of experts deploring the "white poison", it was not until Leclaire developed his industrial process for zinc white in the 1840s that this public health issue gained strong visibility.

Once again, after the mid-century flurry of public action to establish a ban (1849–1852), a second period of heavy silence began, lasting a half-century. Three major factors were behind this period of waning controversy. First, the white lead industry formed a powerful lobby group under the Second Empire. Mainly operating behind the scenes, this pro-lead coalition became very active in the mid-19th century to fend off the government's attempts to ban white lead. French white lead manufacturers were few in number (around a dozen nationwide), but were very concentrated geographically (Lille, Paris and Tours). These industrial powerhouses were strengthened with cross-shareholdings and often backed by family ties — especially with the powerful textile industry in Northern France — and were well-organized to exert their influence on decisive regional and national stakeholders, strong enough to maintain white lead production despite opposition to the material.

The Third Republic opportunistically supported the white lead industry, thanks to the efforts of Charles Expert-Bezançon, who was the leading white lead industrialist in Paris, an influential member of French industrialist circles and senator for Paris *département* (1900–1909). He was a spokesman, or even a tribune, for the powerful white lead lobbying group over several decades, which invested considerable efforts, in publications and public events, to foster doubts about the toxicity of white lead and the capacity of zinc white to be an effective substitute. Refusing to consider the rosters of sick workers treated in hospitals, challenging any attempts to quantify the lead poisoning epidemic, denying that white lead could be dangerous when it was handled with all the "desirable" precautions, the white lead industrialists echoed hygienist arguments that workers were negligent in protecting themselves. They also skillfully fueled confusion about other toxic materials or other sectors responsible for lead poisoning that did not use white lead (printing, plumbing, battery manufacturing, etc.). Boasting their industry's technical advances since the mid-19th century, they endeavored to cast doubt on the validity and the hidden motives of the campaign against white lead, whose assertions were considered baseless and in no way related to epidemiological reality [12].[4] From a technical standpoint, the white lead industry sought to prove that lead-based paint was superior to its zinc-based rival by shifting the debate to industry experts and away from physicians or the general public, considered uninformed about opacity, ease of application, drying and resistance of the paint over time. From a political standpoint, the approach repeatedly sparked suspicion about the true motivations of the opponents to white lead, accusing physicians, unionists and politicians of being corrupt and "anti-patriotic," allegedly being paid by the Belgian or German zinc industry to destroy the French white lead industry [13].

Lastly, from a medical standpoint, the white lead industry attempted to manipulate the nascent clinical knowledge about lead poisoning. Substantial research focused on lead poisoning after 1830, but the findings were largely limited to a scientific audience. Following Tanquerel des Planches (1834, 1839) and Grisolle (1835), clinical knowledge in the late 19th century specified the many symptoms and pathologies all connected to lead — designated without a doubt as a violent poison. However, this clinical knowledge and the medical controversies were inherently technical, remaining somewhat invisible to the general public. Only at the turn of the 20th century did the medical and political fields partly become synchronized, notably through networks of republicans and Free Masons,

---

[4]E.g., Fleury (1905).

such as Paul Brouardel, a friend and follower of Louis Pasteur and professor of forensic medicine as well as an eminent member of the Academy of Medicine. During the 1900–1910 decade, medical expertise on occupational lead poisoning found an audience outside the hospital sphere: in the political arena of the Third Republic, the interconnected but contradictory demands of personal freedom and social justice brought scientific experts back into the political spotlight that they had partially deserted during the 19th century [14].

The labor movement in the latter half of the 19th century was not mature enough to publicize the health issues of industrial poisons and to demand about suitable occupational health conditions in the chemicals industry. There were no official labor unions in white lead factories. Over the previous decades, hygienist journals had published a fairly large number of articles about arsenic, lead and mercury poisoning, but the topic was nearly ignored by the nascent labor movement, which was mostly focused on achieving political and doctrinal unity, and emphasized fundamental demands for higher wages and reduced working hours. It was not until 1900, when house painters began to organize as a trade (they were the main users of white lead) that labor unions started to refer to lead poisoning as a "cause." It remained nevertheless a secondary topic, quickly set aside when not considered useful for building a unified labor movement [15]. Ultimately, the converging labor and medical movements shifted the issue of lead paint poisoning into the public eye. No longer confined to the sidelines after a half-century, white lead entered the spotlight thanks to the convergence of the labor movement, scientists speaking out about public health, social policy being promoted by government leaders and the press's new-found interest in publicizing this cause, designing a new "regime of perceptibility" [16].[5] This convergence of influences gave rise to new national rules, with the 1909 act that banned the use of white lead in building paint and attempted to eliminate the risk for house painters. This was only partially successful, however, because the ban,

---

[5] Michelle Murphy uses the term "regime of perceptibility" to refer to this kind of configuration that is favorable to bringing a health issue into the public debate. See Murphy (2006).

scheduled to take effect on January 1, 1915, was overtaken by the advent of World War I.

After the Armistice, the efforts to fuel doubts about zinc white as a possible substitute for white lead reappeared during the heated debates at the 3rd International Labor Conference, starting the international career of the product. The recently-founded ILO (1919) quickly addressed the issue as a sign of its determination to set standards for sanitary working conditions in order to draft an International Convention banning white lead. However, even a century and a half after Guyton de Morveau's demonstrations, the qualities of the two materials were again addressed for several weeks in Geneva, suggesting that a scientific debate was still legitimate and unsettled. As Albert Thomas noted at the time, the advocates of white lead insisted on "undermining the authority of the scientific demonstrations on which [the convention] was based! It is therefore important [...] that the experts [...] defend the truth against those who wish to adulterate it" [17].[6]

4. When Poison Enters the Marketplace

Endorsing the liberal anti-interventionist credo that the marketplace should act as the "site of veridiction" [18], the advocates of white lead opposed government intervention for the sake of open economic competition, which they claimed revealed its true value and thus should be the sole determinant: "When the railways were built, the stage coaches disappeared; they died a timely death. If zinc white is truly superior to white lead, it will kill us in the marketplace, but the government should not intervene" [19]. These were the words of Expert-Bezançon, in his February 1903 deposition to the parliamentary committee examining the bill for banning lead-based pigments in paint. Among the industrialists, the debate was essentially about political economy and free enterprise, and the denied right of the State to remove a product due to drawbacks extrinsic to the market.

---

[6] Speech by the Director of the ILO (1923).

However, several factors disrupted the competitive landscape shaped by the invention of the manufactured zinc white as an alternative product to the poisonous white lead. White lead's longer presence in the paint market was a major advantage for the material. It had been a luxury good, before being manufactured on an industrial scale inundating the Western European markets for paint for artists and decorations. When Leclaire's zinc white entered the market as of 1845, white lead held a dominant position, having reigned supreme in the building paint markets under its many avatars: Venetian ceruse, Meudon, Spanish, Champagne or Troyes whites, all derivatives of the original pigment, named according to the proportion of white lead content. Zinc white would have to overtake white lead. Beginning in the 1840s, Belgian firm Vieille-Montagne — leader in the European zinc market — implemented a clever strategy to win over the paint market. It acquired and filed for patents, built subsidiaries, carried out various experiments, made verbal and secret agreements to buy out competitors, all in an attempt to defeat white lead. However, the white lead industry enjoyed substantial private and political support under the Second Empire, even though the French government put a legal — but never implemented — ban in public works projects in 1849. Despite the financial firepower of the Vieille-Montagne, which manufactured and sold zinc white in the French, Belgian and German markets, zinc white struggled to challenge white lead in the paint market until the 1920s. Similar in price to its competitor, zinc-based paint nevertheless required painters to work differently: despite its harmlessness and because of this unwanted technical turn in the workers' know-how, zinc white was rejected by most paint professionals.

In the end, however, the health argument won out — at least in public opinion. The health argument or, again, a matter of price? While estimates of the number of occupational lead poisoning victims had been imprecise and very incomplete, by the mid-19th century, several publications revealed the financial cost of these diseases for society, in addition to the images of people suffering. Figures showed the growing number of white lead workers and painters suffering from poisoning, giving them a new-found visibility in hospital wards: 948 patients hospitalized in Paris for lead poisoning over a period of 942 days (from September 1849 to March

1852), costing the Paris hospital administration some 19,000 days of care to lead poisoned patients [20]. Crippled men with arm and leg paralysis unable to work and depending on charity for life, weak and disabled children born to lead-poisoned mothers, families forced into penury by the early death of fathers: it eventually appeared that including the health and social cost of chronic lead poisoning in the calculation made white lead much more expensive than its substitute. In the latter half of the 19th century, references to the cost of treating lead poisoning victims and taking care of the disabled for life became a leitmotiv of those arguing that zinc white should be preferred "in the interests of humanity *and the state*".[7]

Ultimately, the biography of white lead as a serial killer raises the basic question of competition in the marketplace between two materials whose qualities and value were built under different criteria, along different timelines. On the one hand, technical knowledge determined the chemical makeup, utility and price of each material. On the other hand, medical knowledge and expertise assessed each material's harmfulness for human health. Thus, the purely industrial and commercial motives were overshadowed by the health issues, and the resulting regulatory standards impacted the markets by changing the notion of quality and thus also market conditions. Consumers' "choice" had little to do with defining quality [21]. White lead's commercial resilience was attributable less to its supposed intrinsic technical qualities (ease of application and drying, opacity, weather resistance, etc.) than to the qualities attributed to it in a specific historical context: its centuries-old reputation, national production capacity and "traditions" in the painter's trade, all underpinned by an unprecedented public lobbying campaign and much concerted effort by stakeholders. Conversely, during the periods when the health argument prevailed as a driver of public action (namely, late 18th century, 1849–1852 and 1900–1930s), the material's alleged merits were completely overshadowed by its harmfulness that somehow swept away the economic conflict by asserting that human life was more important than the market. As Justin Godart, French delegate at the ILO, claimed in 1921:

"Faced with all these private interests defending themselves [...], we must consider a higher interest: human interest. I believe that by

---

[7] *Ibid.*: 5. Italics in the text.

protecting human life [...], we serve the general economy and commonwealth more usefully than by allowing white lead to be used, *even if* it is an established fact that it provides longer protection to the materials it covers, *even if* its cost is lower" [23].[8]

In these circumstances, public health and free competition seem two irreconcilable facets of economic growth. Nevertheless, a transnational system of standards was crafted in an attempt to reconcile these two values in favor of public health. Widely unsuccessfully.

5. Conclusion

The long controversy between lead and zinc whites exemplifies the strong interconnections between economic, social, technical and political factors that determine the life and the identity of materials. Turning a useful and common material such as white lead into an industrial poison was a complex process involving scientific research (in medicine and chemistry), technical debates, economic rivalries and political sparring. For more than a century, self-proclaimed experts who would later be known as "merchants of doubt" [24] maintained a fog of social and political doubt about the dangers of white lead and the innocuousness of zinc white. Even though the latter had been scientifically established by the end of the 18th century, they did not hinder the continued commercial dominance of white lead. In order to disentangle the mesh of converging interests interweaving the alliance between white lead and industrial societies, scientific evidence may be necessary but it is by no means sufficient. Industrial and social stakeholders use mechanisms for manufacturing ignorance or, at the very least, perpetuating doubt about medical and chemical expertise thus creating scientific confusion and thus breaking the market rules. Powerful public and private actors thus participated in the elaboration of a regulation of the toxic which, while relying on the elaboration of a protective norm for people, permitted the resilience of the material and the social acceptance of a risk inherent in the industrial society of 'progress'.

---

[8] Report of the 22nd session of the International Labour Organization (1921 November 17), Geneva. Emphasis added.

References

1. Homburgand, E. and De Vlieger, J. H. (1996). A victory of practice over science: The unsuccessful modernization of the Dutch White lead industry (1780–1865). *History and Technology*. 13, 33–52.
2. Rainhorn, J. (2019). *Blanc de plomb. Histoire d'un poison légal*. Paris: Presses de Sciences Po.
3. Warren, C. (2001). *Brush with Death. A Social History of Lead Poisoning*. Baltimore: Johns-Hopkins University Press; Bartrip, P. (2002). *The Home Office and the Dangerous Trades. Regulating Occupational Disease in Victorian and Edwardian Britain*. Amsterdam: Rodopi; Malone, C. (2003). *Women's Bodies and Dangerous Trades in England, 1880–1914*. Woodbridge: The Boydell Press.
4. Ramazzini, B. (1700). *De Morbis Artificum Diatriba*. Padua.
5. Fourcroy, Berthollet, Vauquelin, Report to the National Institute (1808), quoted in Jouffroy d'Abbans, A. (1860). *Dictionnaire des inventions et découvertes anciennes et modernes dans les sciences, les arts et l'industrie*. Paris: Migne. pp. 1406.
6. Subsequently called "Vieille-Montagne". See Peters, A. (2016). *La Vieille-Montagne (1806–1873). Innovation et mutations dans l'industrie du zinc*. Liège: Éditions de la Province de Liège.
7. Robert, C. F. (1879). *Biographie d'un homme utile. Leclaire, peintre en bâtiments*. Paris: Sandoz et Fischbacher. p. 39.
8. Chevallier, A. (1849 January). *Rapport fait à la Société d'encouragement pour l'industrie nationale au nom du Comité des arts chimiques, sur la substitution du blanc de zinc au blanc de plomb et que couleurs à base de plomb et de cuivre, par Leclaire*. Paris, Bouchard-Huzard.
9. Proctor, R. and Schiebinger, L. (2008). *Agnotology. The Making & Unmaking of Ignorance*. Stanford: Stanford University Press; Proctor, R. and Holocaust, G. (2012). *Origins of the Cigarette Catastrophe and the Case for Abolition*. Berkeley: University of California Press; Boudia, S. and Jas, N. (Eds.) (2013). *Powerless Science? Science and Politics in a Toxic World*. New York: Berghan Books.
10. Le Roux, T. (2011). L'effacement du corps de l'ouvrier. La santé au travail lors de la première industrialisation de Paris. *Le mouvement social*. 234(1), 103–119.
11. Le Roux, T. (2011). *Le laboratoire des pollutions industrielles, 1770–1830*. Paris: Albin Michel.

12. Fleury, P. (1905). *Blanc de zinc et blanc de céruse, historique complet de la question, réfutation du rapport de la commission législative, critique raisonnée du projet de loi voté à la Chambre des Députés le 30 juin 1903.* Coulommiers: Imprimerie Brodard. p. 74.

13. Dubreuil, M. (1905). *Blanc de zinc contre blanc de céruse. Trente ans de céruse, ou l'avis d'un pinceau.* Paris: L'Alliance latine.

14. Topalov, C. (Ed.) (1999). *Laboratoires du nouveau siècle. La nébuleuse réformatrice et ses réseaux en France (1880–1914).* Paris: EHESS.

15. Rainhorn, J. (2012). Workers against lead paint: How local practices go against the prevailing union strategy (France, at the beginning of the 20th c.). In Bluma, L. and Uhl, K. (Eds.) *Kontrollierte Arbeit — Disziplinierte Körper? Zur Sozial — und Kulturgeschichte der Industriearbeit im 19. und 20. Jahrhundert.* Bielefield: Transcript Verlag. pp. 137–161.

16. Murphy, M. (2006). *Sick Building Syndrome and the Problem of Uncertainty: Environmental Politics, Technoscience, and Women Workers.* Durham: Duke University Press.

17. Thomas, A. Manifestation contre la céruse. (1923). *Organisée par la Ligue des Sociétés de la Croix-Rouge, le 3 mars 1923 dans le grand amphithéâtre de la Sorbonne.* Paris. p. 29, Imprimerie Deshayes.

18. Foucault, M. (2008). *The Birth of Biopolitics. Lectures at the Collège de France (1978–1979).* New York: Palgrave McMillan. p. 32.

19. Breton, J. L. (1903). Rapport supplémentaire fait au nom de la commission chargée d'examiner le projet de loi sur l'emploi des composés du plomb dans les travaux de la peinture en bâtiments. *Chambre des Députés.* annexe XI, 245.

20. Richelot, G. (1852). *De la substitution du blanc de zinc au blanc de plomb dans l'industrie et les arts.* Paris: Félix Malteste et Cie. p. 13.

21. Stanziani, A. (2008). La définition de la qualité des produits dans une économie de marché. *L'Économie politique.* 37(1), 95–112.

Chapter 11

Asbestos

Emmanuel Henry

Université Paris-Dauphine, PSL University,
75016 Paris, France

The generic term "asbestos" actually refers to different types of silicate minerals, which all share the common characteristic of being fibrous. To the touch, such a rock quickly disintegrates into extremely fine fibers, the different characteristics and qualities of which have made it an extremely popular material in the 20th and 21st centuries. Beyond this direct observation, these fibers themselves comprise microscopic fibers invisible to the naked eye that can penetrate the human body — hence the dangers associated with asbestos. The health effects linked to the inhalation of this material were highlighted early on, so much so that asbestos became one of the most controversial materials of the 20th century, regularly prompting large mobilizations from the 1970s. Although the quantity of asbestos used in the world very recently started to slowly decline, it remains a particularly clear example of the continued use of a product even when its dangers have been extremely well known and confirmed by scientific work for several decades. While many studies have highlighted ignorance in the production of knowledge and regulations, [1] asbestos actually strikingly exemplifies a case where scientific knowledge exists but is powerless against the weight of economic interest in promoting and maintaining its use [2]. Just as asbestos is a fibrous material that can form different layers, so too does its biography reflect a layering of various characteristics and events. These different dimensions range from its physical–chemical properties that have made it so successful, to the scientific discoveries about the risks associated with its inhalation and the forms of regulation that have allowed for its continued or even encouraged use.

1. A "Magic Mineral"

Asbestos is a naturally occurring mineral in certain geological configurations. Its mining developed primarily in four countries: Canada, South Africa, Russia and Brazil. Although asbestos had been used on occasion for a very long time, its industrial use started to develop from the early 20th century and accelerated after the Second World War. The infatuation with asbestos is owed to a range of combined characteristics that make it a unique material, both extremely flexible and extremely strong. In addition to the flexibility associated with its fibrous nature, asbestos is an excellent heat and electricity insulator, and is also fire and flame resistant. Asbestos thus has many applications in the construction, textile, industrial and shipbuilding sectors. It has been used in flocking to insulate certain buildings and boats and make them fire-resistant. It has also been used to insulate certain fabrics and make them flame-resistant, firefighter suits for example. More recently and on a more massive scale, asbestos has been used extensively with cement, either to make pipes or to make asbestos–cement slabs used as roofing, or in certain materials to reinforce buildings. The low cost of asbestos explains the extremely rapid proliferation of its applications and the scarcity of research into possible substitutes.

In quantitative terms, global asbestos consumption grew until 1980, reaching a peak of 4.6 million metric tons per year [3]. It has since slowly declined, stabilizing around a still extremely high annual consumption of 2 million tons. Since the early 2000s, countries like China and Russia have been consuming over 500,000 metric tons of asbestos per year, a trend that only started to show signs of slowing down from 2015. Yet, measuring annual flows of asbestos consumption paints a misleading picture of the dangers of this mineral. As it is virtually indestructible, its accumulation should rather be analyzed in terms of its accretion [4]. Worldwide, cumulatively, over 200 million metric tons of asbestos have been produced since 1900. It is present in buildings in every city and in many agricultural buildings. Asbestos extracted from rock and concentrated or integrated with other materials is becoming a ubiquitous poison in our environment that threatens our existence and will continue to do so in the coming decades.

2. Discovery and Confirmation of the Dangers of Asbestos

The harmfulness of asbestos to human bodies is now well documented. Yet, the history of the discovery of the health effects of asbestos exposure is emblematic of the interplay between environmental and occupational health.

Asbestos exemplifies a process common to many other occupational and environmental toxicants, which sheds light on the forgotten or at least opaque interdependence between occupational and public exposure [5]. As with many other products present in the environment and in the workplace, the alerts resulting from occupational exposure served to draw attention to the asbestos risk for the population as a whole. The dangers of asbestos were established through several studies on cohorts of workers. In the early 20th century, research on weaving mills, such as the study carried out by a French labor inspector in 1906 [6], drew attention to asbestosis. Scientific research then confirmed the carcinogenic nature of asbestos, associated with both lung cancer and mesothelioma, between the 1930s and 1960s. Thus, various studies were needed, among Canadian and South African miners, workers in asbestos processing plants (textile and cement) and insulation workers [7].

Workers' frontline position in the discovery of the health effects of a toxin is standard in the case of environmental toxins. The dangerousness of a toxin at the high doses to which workers are exposed is determined, and then draws attention to the dangers of lower exposure for the general population. A similar situation was observed with diesel particulate emissions, where adverse effects ascertained through studies of miners and truck drivers confirmed the hazard to the general population exposed to lower doses [8–10]. As with diesel in 2012, the classification of asbestos as a carcinogen by the International Agency for Research on Cancer (IARC) in 1977 played a central role in sounding the alert internationally [11]. From that point onward, the carcinogenicity of asbestos was definitively confirmed in the eyes of different scientific communities and received at least occasional attention in wider public spheres. The 1970s thus saw the issue of asbestos-related risks receive more attention in society, whether through journalistic investigations and lawsuits in the United States [12,13], or social mobilization in France [14].

Yet, this greater visibility of the dangers of asbestos did not translate into policy change regarding this material. On the contrary, over 80% of all asbestos came into use *after 1960*, that is, after the scientific confirmation of its carcinogenicity and confirmation of its serious risks to public health [15]. Although regulation of the use of asbestos developed from the 1970s in some countries (in the United States and the European Community in particular), only in the 1990s did the first bans on asbestos enter into effect, in a few Northern European countries. The European Community did discuss a directive banning asbestos in the early 1990s, but it was fiercely and successfully opposed by France, which delayed an EU-wide ban on asbestos by more than 20 years.

There is therefore no correlation between the scientific confirmation of the danger associated with this material and the far more recent decline in its use. Nowadays, the use of asbestos is banned in 50 countries only, including the 28 Member States of the European Union. It is still authorized in India, China, Russia, Brazil and Indonesia, to name but a few of the main asbestos-using countries. Nor is asbestos banned in the United States, although its use there is limited [16]. What dynamics could explain this situation, including within the scientific sphere and among the actors having to make decisions about this product?

3. Industrial Manipulations and the Dynamics of the Scientific Field

Several dynamics have converged to the continued use of this material over several decades after the demonstration of its damages on human health. Some of these have already been observed in the case of other hazardous materials, while others are more specific to asbestos or are sometimes more common to the toxic substances primarily affecting workers.

First of all, asbestos firms have mobilized in specific ways to steer the dynamics of scientific research in a direction that favors their interests. This way of ignorance production is quite similar to that observed in the case of the tobacco industry [17] and of other occupational toxicants such as silica [18], lead or vinyl chloride [19].

In the case of asbestos, the main strategies have involved differentiating between more or less dangerous types. For example, industry has drawn on scientific studies to support the view that chrysotile, the most commonly used type of asbestos, is less dangerous than other types [20,21]. This distinction has been used to justify the claim that exposure to chrysotile is safe below a certain threshold [15]. The general idea behind this strategy is to defend the idea that it is possible to maintain a "controlled use" of asbestos, at least in workplaces [22].

The direct intervention of the asbestos industry in promoting these controversies is well documented, whether it be the links between Richard Doll and some manufacturers such as Turner & Newall [23] or the range of research funded by the various employer associations of the asbestos industry [15]. It is important, however, to stress that these investments were made all the more effective by the fact that they did not directly confront the research conducted in the scientific sphere itself, but steered them in specific directions. Creating new controversies around the distinction between categories of asbestos, or seeking a threshold of exposure below which asbestos would be less or no longer dangerous, modified the agenda of the scientific community. Other researchers were thus encouraged to engage in these new controversies to facilitate future publications even without any financial incentive. Therefore, the industry's influence on scientists should not only be analyzed as gross manipulation by economic actors (which it is sometimes also). But by limiting themselves to encouraging and reinforcing dynamics internal to the scientific field, industrialists could actually ensure that the effects of their investments in certain scientific teams were amplified [24,25].

Beyond these production-of-ignorance approaches similar to those developed for other toxins, asbestos is a characteristic example of an old toxicant that is scientifically extremely well known — despite some controversies. Yet, knowledge of this toxicant has remained in scientific circles without being taken into account by public decision-makers. The case of asbestos reveals the very strong autonomy of social, economic and political dynamics, from scientific ones. The scientifically proven carcinogenicity of asbestos did not prevent its continued use in the decades following this discovery. Rather than ignorance or undone science [26], this situation is best captured by the concept of *unseen science*, as has been argued in the case of fluorinated compounds [27]. The regulation of asbestos appears to be fairly characteristic of a situation of impermeability between scientific spaces, in which the confirmation of the dangerousness of asbestos circulates, and regulatory spaces, where the controversial dimension is more visible [28–30]. Even though the carcinogenicity of asbestos has never been contested as such by scientists, the controversies that have circulated in the scientific sphere have made it difficult for readers outside that space to get a clear idea of the extent of the dangers of asbestos. This difficulty in assessing hazards and risks has been further exacerbated by the fact that few studies have been carried out to estimate the number of illnesses and deaths associated with asbestos exposure, such as the one that helped to alert European decision-makers in the mid-1990s, based on a quantification of the epidemic in Britain [31]. This non-transmission of knowledge from the scientific to the regulatory spheres has strongly contributed to making asbestos-related pathologies a largely invisible epidemic for policy-makers.

4. Ignorance and Short Fibers

Asbestos also illustrates the forms of ignorance produced by scientific tools and the resulting political and social conventions. For a very long time, the microscopes used to count asbestos fibers could measure only those at least 5 μm in length. To be more precise, the so-called "WHO fiber" is an asbestos fiber longer than 5 μm, with a diameter below or equal to 3 μm and a length-to-diameter ratio of less than 3 [32]. This fiber size was the result of a consensus negotiated in the 1960s and endorsed by the World Health Organization (WHO), to standardize dust measurements worldwide [33]. This standard had no health justification, and was essentially dictated by the type of microscope used (the phase-contrast optical microscope). It merely corresponded to the fibers that were technically measurable in a systematic way at the time.

The development of the electron microscope (analytical transmission electron microscope, ATEM) gradually changed this state of affairs. By making it possible to measure finer and shorter asbestos fibers, it challenged certain compromises that had previously governed the dust control measures. The agreement on the standardization of the size of asbestos fibers allowed for compromises to be reached around the use of asbestos. At the same time, it has produced ignorance, as a range of questions regarding the health effects of certain types of fibers were never raised. Fine or short asbestos fibers could not be measured, they had no social or scientific existence, and their toxicity was not being investigated in any way. Due to a semantic confusion, measured fibers, which were subject to regulation, have been conflated with dangerous fibers, as noted by the researcher who steered the scientific review that led to re-evaluation of the issue of fiber size: "The terms 'regulated fiber' and 'pathological active fiber' are often liberally and erroneously interchanged" [34]. Conversely, fibers not counted in the various international regulations applicable could appear to be less dangerous (sometimes as a result of simple common-sense reasoning causing actors who were not well informed on these issues to make the logical assumption that if these fibers were not regulated, that meant they were not dangerous).

When these fibers began to take on a social existence, that is, once they could be observed by electron microscope, industrial actors in the sector were quick to exploit their prior non-existence and the resulting ignorance of their toxicity. Just as they had done in the 1970s and 1980s with the search for a possible threshold of asbestos innocuousness and the differentiation of asbestos types, in the late 1990s–early 2000s industrial actors grabbed the opportunity to use ignorance to highlight the uncertainty surrounding the toxicity of short asbestos fibers (SAF). A controversy developed, heightened by the fact that workers in certain branches of industry, notably automobile repair, where they handled brakes containing asbestos, may have been mainly exposed to short fibers. As SAF had only recently come into social existence, their level of toxicity remained uncertain. This residual uncertainty was soon framed as a lack of scientific evidence, or even an absence of risk altogether [35,36].

In France, the Ministry of Labor thus requested an expert report from AFSSET (*Agence française de sécurité sanitaire de l'environnement et du travail*, the French environmental and occupational health safety agency) on fine and short asbestos fibers [37]. In its conclusion, the AFSSET report reached a position that also clearly illustrated experts' dependence on the organization of the scientific research. Due to the lack of incentives and funding for research in these areas, few scientific data were available regarding the toxicity of SAF, and the expert agency could only go so far as to state that such toxicity could not be ruled out, while presuming that it would be lower than that of long fibers: "While the direct or indirect toxicity [...] of SAF remains difficult to assess, it cannot be ruled out. Should SAF be toxic, this toxicity would certainly be lower than that of long fibers, though no weighting can be defined at this time" [33].

5. Confining the Risk to Occupational Populations

The perpetuation of the use of asbestos is rooted in the belief that the epidemic is limited to workers, and therefore controllable and limited. This belief has of course been encouraged and reinforced by employers, who have a vested interest in developing the use of this material [22]. But it has spread all the more easily due to the fact that it also supports the interests of different categories of actors, particularly political actors, who had a vested interest in encouraging the development of an extremely profitable economic activity [38]. The use of asbestos has thus developed very easily because its health effects seem distant and localized to the people in charge of decisions about their regulation [39]. Thus, aside from the bans which, as we have seen, were enforced late and applied in a limited number of countries only, the regulation of asbestos has differed not only from one country to the next but also within each country, based on the circumstances of exposure and the categories of actors threatened or exposed to the fiber.

Prominent among the instruments that have sustained this belief in the controlled use of asbestos was the regulation of occupational exposures on the basis of thresholds or limit values: from the 1970s, various limit values governed occupational exposure to asbestos. Yet, as the carcinogenic nature of asbestos has been established since this period, it has been known that exposure, even at lower levels, can lead to the development of associated pathologies. This is very clearly stated by the experts gathered at the International Labor Organization (ILO). "The experts recommended that, in the present state of knowledge, the 2 fibres/ml time-weighted standard which had been adopted by some member States should be regarded as an interim target concentration for the prevention of risk to the health of asbestos workers. It was recognized that this standard related to the fibrogenic effects and not to the carcinogenic effects for which no standards exist at the present time" [40].

As scientists have known for decades, no threshold limit values guarantee the absence of cancer risk. What the thresholds produce, instead, is a political solution, one that establishes a tradeoff between public health and economic growth. The limit values for asbestos express a tacit agreement to place a certain number of peoples' health and lives at risk in exchange for the protection of industries that manufacture and use asbestos. The tacit agreement is built on and reinforces the assumption that "safe" occupational exposures exist, even as those values maintain significant levels of risk among some groups of workers and legitimize occupational exposure that will result decades later in a growing number of asbestos-related diseases [41].

6. Illnesses and Deaths Primarily Affecting Workers

Asbestos primarily affects victims contaminated through their work, first and foremost because although exposure levels have gradually declined over time, they have overall remained far higher in occupational settings. These levels reveal not so much a differentiation between countries, as a social and occupational mapping of the epidemic. This differentiated exposure also sheds light on a timelag in exposures, particularly in the case of countries that have banned or severely restricted the use of asbestos and have seen a gradual drop in certain highly exposed occupational groups. The group that is both the most affected and historically the first to have been affected comprises miners and workers in asbestos processing plants. Whether in textile factories or those manufacturing asbestos–cement slabs, asbestos exposure is particularly high, leading to epidemics of asbestosis followed by cancer affecting an extremely large proportion of exposed workers. Workers in charge of removing asbestos in buildings are now those most likely to be exposed to asbestos at very high levels.

The second circle of exposed individuals are the workers of companies using asbestos for an extremely wide range of activities, such as insulation companies (particularly those spraying asbestos), shipyard workers and, more broadly, individuals working in confined environments exposed to asbestos. Although levels of exposure in these settings are generally slightly lower than for the first category of workers most affected, they remain quite high and affect a larger population of workers. Since the 1990s, countries that have restricted the use of asbestos have been primarily concerned with occupations exposed to asbestos on a less regular basis and that are not specifically attentive to this risk. This includes in particular all occupations involved in the finishing of buildings (roofers, plumbers, electricians, etc.). Since the 1990s, construction workers have been the occupational group most affected by asbestos-related diseases in Western Europe [31]. While exposure is more sporadic for these groups, it may attain high peaks, the effects of which significantly increase cancer risks. Concern about these occupational groups is more recent but has become particularly acute, insofar as these occupational groups represent much larger populations than those initially identified (in France, over 1.2 million construction workers face exposure risks; 6 million in the United States). As a result, industrial companies manufacturing or using asbestos have been submitted to many lawsuits. By 2006, ABB — a Swiss-Swedish multinational corporation that acquired and took control of the US-based Combustion Engineering Company in 1991 and assumed responsibility for the damages caused by the latter manufacture — had to deal with so many compensation claims with regards to mesothelioma lawsuits that in 2006 the company set up a trust fund of about $1.5 billion to settle pending and future claims.

In addition to occupational victims, other groups of victims are also impacted by asbestos exposure, albeit to a lesser extent. The first group comprises individuals subject to so-called para-occupational exposure. Such exposure affects people living with or near asbestos workers — their families, relatives and neighbors. One can easily imagine a factory worker's spouse inhaling asbestos from the worker's overalls while doing the family laundry. A case that became widely known in France in the 1990s involved a young butcher who died of mesothelioma at age 28. He had been exposed to asbestos as a child, playing with the neighbors' children whose father worked at an Eternit fiber cement factory in northern France [38].

Another group of asbestos victims are people living in places where asbestos is mined or processed, where the fibers circulate at high levels throughout the natural environment. In a few exceptional cases, such as in New Caledonia, Spain and various counties in California, the source of asbestos may be a natural outcropping of the mineral linked to certain geological features of the soil.

Overall, however, the vast majority of asbestos victims are exposed to it in an occupational setting. According to the most recent global estimates, asbestos kills about 255,000 people every year, mainly as a result of occupational exposure (in an estimated 233,000 cases) [42]. This figure is impressive both because of its magnitude and given the social silence surrounding it. As the mobilizations at national level to make asbestos a key public issue show, the mainly occupational nature of the epidemic makes it difficult to fully appreciate the health consequences of asbestos exposure, in turn making it difficult to enforce a comprehensive ban.

7. Conclusion

It is impossible to accurately predict the future of this epidemic. The number of asbestos-related diseases is highly correlated with asbestos use, with the timeframe of disease progression entailing a 30- to 40-year lag [31]. While many journalists and analysts consider that the asbestos epidemic is behind us, the harsh reality is that millions of people will continue to be exposed to asbestos for decades to come. In the United Kingdom, for example, asbestos-related diseases will continue to increase until about 2020, even though imports of asbestos have been declining since the late 1970s [43]. In France, which enforced a total ban on asbestos use in 1997, one can expect a very slow decrease in the annual number of asbestos-related diseases beginning around 2030. Ultimately, even if asbestos were banned worldwide tomorrow, the epidemic of asbestos-related diseases would persist at least through the next decades.

References

1. Gross, M. and McGoey, L. (Eds.) (2015). *Routledge International Handbook of Ignorance Studies*. Abingdon, New York: Routledge. p. 426.

2. Boudia, S. and Jas, N. (Eds.) (2014). *Powerless Science? Science and Politics in a Toxic World*. New York, Oxford: Berghahn Books. p. 280.

3. Virta, R. L. (2006). *Worldwide Asbestos Supply and Consumption Trends from 1900 through 2003*. Reston: U.S. Geological Survey. p. 80.

4. Boudia, S., Creager, A. N. H. and Frickel, S. *et al.* (2021). *Residues: Thinking through Chemical Environments*. New Brunswick, NJ: Rutgers University Press.

5. Sellers, C. C. (1997). *Hazards of the Job: From Industrial Disease to Environmental Health Science*. Chapel Hill: University of North Carolina Press. p. 331.

6. Auribault, D. (1906). Note sur l'hygiène et la sécurité des ouvriers dans les filatures et tissages d'amiante. *Bulletin de l'inspection du travail*. 120–132.

7. Castleman, B. I. and Berger, S. L. (2005). *Asbestos: Medical and Legal Aspects*. New York, NY: Aspen Publishers. p. 894.

8. Attfield, M. D., Schleiff, P. L. and Lubin, J. H. *et al.* (2012). The Diesel Exhaust in Miners study: A cohort mortality study with emphasis on lung cancer. *Journal of the National Cancer Institute*. 104, 869–883.

9. Silverman, D. T., Samanic, C. M. and Lubin, J. H. *et al.* (2012). The Diesel Exhaust in Miners study: A nested case-control study of lung cancer and diesel exhaust. *Journal of the National Cancer Institute*. 104, 855–868.

10. Garshick, E., Laden, F., Hart, J. E., Davis, M. E., Eisen, E. A. and Smith, T. J. (2012). Lung cancer and elemental carbon exposure in trucking industry workers. *Environmental Health Perspectives*. 120, 1301–1306.

11. International Agency for Research on Cancer. (1977). *Monographs on the Evaluation of Carcinogenic Risk of Chemicals to Man. Asbestos*. Vol. 14. Lyon: IARC.

12. Brodeur, P. (1974). *Expendable Americans*. New York: Viking Press. p. 274.

13. Brodeur, P. (1985). *Outrageous Misconduct: The Asbestos Industry on Trial*. New York: Pantheon Books. p. 374.

14. Collectif intersyndical sécurité des universités Jussieu CFDT CGT FEN. (1977). *Danger! Amiante*. Paris: François Maspéro. p. 423.

15. McCulloch, J. and Tweedale, G. (2008). *Defending the Indefensible: The Global Asbestos Industry and its Fight for Survival*. New York: Oxford University Press. p. 325.

16. Frank, A. L. and Joshi, T. K. (2014). The global spread of asbestos. *Annals of Global Health*. 80, 257–262.

17. Proctor, R. N. (2011). *Golden Holocaust: Origins of the Cigarette Catastrophe and the Case for Abolition*. Berkeley: University of California Press. p. 737.

18. Rosner, D. and Markowitz, G. E. (1991). *Deadly Dust: Silicosis and the Politics of Occupational Disease in 20th-Century America.* Princeton, NJ: Princeton University Press. p. 229.

19. Markowitz, G. E. and Rosner, D. (2002). *Deceit and Denial: The Deadly Politics of Industrial Pollution.* Berkeley, CA: University of California Press. p. 408.

20. Mossman, B. T., Bignon, J., Corn, M., Seaton, A. and Gee, J. B. (1990). Asbestos: Scientific developments and implications for public policy. *Science.* 247, 294–301.

21. Tweedale, G. and McCulloch, J. (2004). Chrysophiles versus Chrysophobes. The White Asbestos controversy, 1950s–2004. *ISIS.* 95, 239–259.

22. Castleman, B. I. (2003). "Controlled Use" of Asbestos. *International Journal of Occupational and Environmental Health.* 294–298, Vol. 9, No. 3.

23. Tweedale, G. (2007). Hero or Villain? Sir Richard Doll and occupational cancer. *International Journal of Occupational and Environmental Health.* 13, 233–235.

24. Michaels, D. (2008). *Doubt is Their Product: How Industry's Assault on Science Threatens Your Health.* Oxford, New York: Oxford University Press. p. 372.

25. Michaels, D. (2020). *The Triumph of Doubt. Dark Money and the Science of Deception.* Oxford, New York: Oxford University Press. p. 336.

26. Hess, D. J. (2016). *Undone Science. Social Movements, Mobilized Publics, and Industrial Transitions.* Cambridge, MA: MIT Press. p. 258.

27. Richter, L., Cordner, A. and Brown, P. (2018). Non-stick science: Sixty years of research and (in)action on fluorinated compounds. *Social Studies of Science.* 48, 691–714.

28. Braun, L., Greene, A., Manseau, M., Singhal, R., Kisting, S. and Jacobs, N. (2003). Scientific controversy and asbestos: Making disease invisible. *International Journal of Occupational and Environmental Health.* 9, 194–205.

29. Braun, L. and Kisting, S. (2006). Asbestos-related disease in South Africa: The social production of an invisible epidemic. *American Journal of Public Health.* 96, 1386–1396.

30. Henry, E. (2009). Rapports de force et espaces de circulation de discours. Les logiques de redéfinition du problème de l'amiante. In Gilbert, C. and Henry, E. (Eds.) *Comment se construisent les problèmes de santé publique.* Paris: La Découverte. p. 155–174.

31. Peto, J., Hodgson, J. T., Matthews, F. E. and Jones, J. R. (1995). Continuing increase in Mesothelioma mortality in Britain. *The Lancet.* 345, 535–539.

32. WHO/EURO Technical Committee for Monitoring and Evaluating Airborne MMMF. (1985). *Reference Methods for Measuring Airborne Man-made Mineral Fibres (MMMF)*. Copenhagen: WHO (Regional office for Europe).

33. Agence française de sécurité sanitaire de l'environnement et du travail (AFSSET). (2009). *Les fibres courtes et les fibres fines d'amiante. Prise en compte du critère dimensionnel pour la caractérisation des risques sanitaires liés à l'inhalation d'amiante. Réévaluation des données toxicologiques, métrologiques et épidémiologiques dans l'optique d'une évaluation des risques sanitaires en population générale et professionnelle*. Maisons-Alfort: AFSSET. p. 394.

34. Dodson, R. F., Atkinson, M. A. and Levin, J. L. (2003). Asbestos fiber length as related to potential pathogenicity: A critical review. *American Journal of Industrial Medicine*. 44, 291–297.

35. Langer, A. M. (2003). Reduction of the biological potential of chrysotile asbestos arising from conditions of service on brake pads. *Regulatory Toxicology and Pharmacology*. 38, 71–77.

36. Goodman, M., Teta, M. J. and Hessel, P. A. *et al.* (2004). Mesothelioma and lung cancer among motor vehicle mechanics: A meta-analysis. *Annals of Occupational Hygiene*. 48, 309–326.

37. Thébaud-Mony, A. (2010). Les fibres courtes d'amiante sont-elles toxiques? Production de connaissances scientifiques et maladies professionnelles. *Sciences Sociales et Santé*. 28, 95–114.

38. Henry, E. (2007). *Amiante: Un scandale improbable. Sociologie d'un problème public*. Rennes: Presses universitaires de Rennes. p. 310.

39. Henry, E. (2011). A new environmental turn? How the environment came to the rescue of occupational health: Asbestos in France, 1970–1995. In Sellers, C. C. and Melling, J. (Ed.) *Dangerous Trade. Industrial Hazards across a Globalizing World*. Philadelphia: Temple University Press. p. 140–152.

40. International Labour Organization. (1974). Report of the meeting of experts on the safe use of asbestos, Geneva, 1973 December 11–18. Geneva: ILO. p. 14.

41. Henry E. (2021), Governing Occupational Exposure Using Thresholds: A Policy Biased Toward Industry, *Science, Technology, & Human Values*, 46, 5, 953–974.

42. Furuya, S., Chimed-Ochir, O., Takahashi, K., David, A. and Takala, J. (2018). Global asbestos disaster. *International Journal of Environmental Research and Public Health*. 15, 1000.

43. Stayner, L., Welch, L. S. and Lemen, R. (2013). The worldwide pandemic of asbestos-related diseases. *Annual Review of Public Health*. 34, 205–216.

Chapter 12

Plutonium

Harry Bernas* and Kate Brown[†]

*CNRS-University Paris-Saclay, France
[†]Massachusetts Institute of Technology

1. Introduction

In 1943, the DuPont Corporation took on the job of building a massive factory for the Manhattan Project to produce the world's first supplies of a new element on the periodic chart — plutonium (Pu).[1] As they drew up plans for nuclear reactors and radio-chemical processing plants to produce plutonium, DuPont officials learned something alarming: "the most extreme health hazard is the product itself" [1]. "It is now estimated," DuPont executive Roger Williams continued, "that five micrograms (0.000005 grams) of the product [plutonium] entering the body through the mouth or nose will constitute a lethal dose." A Manhattan Project medical officer wrote on the margins of this sentence: "Wrong!" With the first birthing pains of plutonium a debate was born over the lethality and consequently liabilities inherent in nuclear industries. Plutonium made not only a novel weapon of war, but its very existence spawned the creation of a new security apparatus to safeguard the "super-poisonous" product. As workers over subsequent decades produced more and more plutonium,

---

[1]There is indirect evidence that it had existed in the early Universe.

it transformed the landscape, science, energy production and human society. Corporate and government officials managed to keep the secret of plutonium's toxicity, but at the cost of militarizing, segregating and compartmentalizing society.

1. Pu at War

A group of scientists led by Glenn Seaborg first synthesized the element at the Berkeley, California, cyclotron on December 14, 1940, via a nuclear reaction producing neptunium-238 so that it would decay to a new element 94. This 94th element was expected to fission more readily than uranium and to produce more neutrons per fission, hence a more effective chain reaction. After irradiating several hundred kilos of natural uranium at the Berkeley and Washington University cyclotrons in 1941, Seaborg, Ed MacMillan and Arthur Wahl at the University of Chicago Metallurgical Laboratory (termed Met Lab) chemically isolated a small sample of plutonium. Seaborg named the new element "plutonium" (in honor of the presumed tenth planet), and chose the symbol Pu (pronounce "pee-uuh") since he anticipated that it would stink. Pu was the only element in the periodic table deliberately made to produce death and destruction.

The secretive Manhattan Project (June 1942–August 1945), directed by Army Corps Engineer General Leslie Groves, centered on creating a nuclear bomb — either from enriched uranium or plutonium. After physicists confirmed that Pu-239 was more fissionable than Uranium-235, the Project scientists focused more resources on the synthesis, chemical isolation, metallurgical transformation and molding of Pu to produce a nuclear bomb.

Having identified a few atoms of Pu, the subsequent mission of several hundred Met Lab scientists and technicians was to figure out how to chemically separate it from uranium on a massive scale. Enrico Fermi and Leo Szilard completed the initial mission when they sustained the world's first nuclear chain reaction in the "CP-1 pile" under the University of Chicago's Stagg Field on December 2, 1942. The second mission involved determining the chemical properties of plutonium. Massive Pu production required a power reactor scaled up a thousand-fold. Plutonium also required a large chemical plant to isolate Pu from the irradiated uranium.

Physicist Eugene Wigner designed a high-power reactor that could produce as much as a kilo of Pu per day. In January 1943, physicist Robert J. Oppenheimer and General Groves set up the Los Alamos Laboratory to test fission characteristics and solve the many technical problems of bomb assembly.

Construction crews began building a pilot X-10 plant in Oak Ridge, Tennessee, in February 1943. There they produced the first significant quantities of Pu in early 1944. That development allowed the first metallurgical studies, which found Pu to be far more complex than any other element in the periodic table. At the same time, DuPont Company officials directed the construction of a much larger production site «W» near Richland, Washington (renamed after the war the Hanford Plutonium Plant). Site W included three production reactors (their number was to reach nine in 1963) and two huge chemical treatment plants on a vast (16,000 ha) area along the Columbia River. The new plant enabled the chemical extraction process of Pu, milked from tons of irradiated uranium, to be scaled up from microgram to kilograms.

Fifty-thousand workers built the plant in a blitz of construction from 1942–1945. The entire Manhattan Project involved over a hundred-thousand workers, the first extensive collaboration between basic and applied researchers and engineers from the entire spectrum of disciplines set to work on a single goal: making a bomb. Pu was at its heart.[2]

Manhattan Project scientists detonated the first plutonium bomb at the Trinity Test Site in July 1945. Fallout from the bomb spread across New Mexico and into the American midwest. A month later, only a week after the Hiroshima U-235 bomb, US army pilots dropped the second Pu bomb, "Fat Man," on the Japanese city of Nagasaki, killing 35,000 people immediately, and leaving years of slow death to tens of thousands more.

After August 1945, Soviets rushed to develop their own bomb. Soviet leaders learned of the new element plutonium through espionage. They

---

[2]The bomb ("Little Boy") dropped on Hiroshima was a U-235 fission weapon. It required over 60 kilos of highly enriched U (HEU): obtaining them was a huge technical feat in itself. But once the HEU was produced, the U-235 bomb — far more rustic than its Pu counterpart — was certain to work. The Pu bomb was a very complex engine in many ways, hence the need to make a first test ("Trinity") in the New Mexico desert.

copied American blueprints for reactors and adopted similar production practices in the years following WWII. In 1946, Project No. 1 commanded a site in the southern Russian Urals and called it "Post Box 40" (later renamed Mayak). Soviets cut costs as they built the Mayak plant with Gulag labor. They hurried into production unconcerned about safety, as building a "nuclear shield" against American aggression was a new wartime front, and people die at the front. Both plutonium plants shared the same feature, which was a high ratio of radioactive waste in relation to plutonium.

Upscaling the Pu chemical extraction process meant separating, every day, plutonium from about a hundred thousand curies (10^{17} becquerels) of radioactive fission fragments — an incredibly dangerous undertaking that produced a massive volume of radioactive waste. In the processing plants, plant operators, who were mostly women in both countries, worked with robotic arms and protective gear behind thick concrete walls and lead-glass windows in ship-sized "canyons." The final product had extraordinary properties. Glenn Seaborg described element 94:

"Plutonium is so unusual as to approach the unbelievable. Under some conditions it can be nearly as hard and brittle as glass; under others, as soft and plastic as lead. It will burn and crumble quickly to powder when heated in air, or slowly disintegrate when kept at room temperature. It undergoes no less than five phase transitions between room temperature and its melting point. … It is unique among all of the chemical elements. And it is fiendishly toxic, even in small amounts" [2].

The toxicity referred to here is twofold: a chemical toxicity comparable to that of other heavy elements such as mercury, and a massive toxicity due to alpha particle radioactivity. A person can handle hardened plutonium, but if they ingest Pu, it reaches vital organs — notably lungs, liver and bones — where it decays.

2. Pu Impact on Human Health

Initially, nothing was known about the biological effects of plutonium and its radioactive by-products. In the US, during the war, researchers injected 50 micrograms of plutonium into 18 unsuspecting human subjects. The previously healthy subjects did not die. Tests showed that their white and red blood cell counts dropped dramatically and their bodies accumulated plutonium with greater efficiency than the bodies of mice. The researchers did no follow up on their secret subjects. Later, family members told of intense pain, weakness, depression, vague complaints and malaise that clouded the subjects' lives afterward.

A few years later, after Soviet engineers built in the Southern Russian Urals a plutonium plant, doctors treated Gulag prisoners suffering from a mysterious nausea and vomiting. The convicts lost weight, had fevers and internal bleeding. The men had dug trenches in high radioactive soil near the processing plant. One of twelve prisoners died. Soviet bosses learned to put prisoners with life sentences on especially dirty jobs. Plant managers worried when young, trained women working in the processing plant began to experience sharp pain in the bones, weakness, chronic fatigue, weight loss and early signs of heart disease. An autopsy of the first young woman who died at age 30 at the Mayak plant showed she had 230 times more plutonium in her bloodstream than « the acceptable norm. » Eight more colleagues died soon after. The cause of death was elusive.

Health physicists in both the US and USSR fixated on gamma rays and external exposures. It took them a long time to understand radioactive exposure from inhalation of dust particles. Eventually, Soviet doctors identified a complex of symptoms they called chronic radiation syndrome (CRS) caused by long-term exposure to low doses of radioactive isotopes. About 2,300 workers at the Mayak plant were diagnosed with CRS. A significant number of these patients died or had cancer before the age of 50.

Americans did not have a similar diagnosis. They tended to give the dirtiest jobs to migrant, minority labor, who left after working a few years. Hanford managers insisted they had no radiation-related deaths at the Hanford plant. Archives show records of two men who died on the job. One independent autopsy indicated the cause of death was radiation poisoning [3].

3. Consequences of Pu Bomb Tests

Rampant testing of plutonium bombs from 1946 to 1998 created a related set of problems connected to nuclear fallout. The atomic bomb delivered radioactivity in two ways. With the blast, Japanese survivors received a single, large dose of gamma rays and neutrons lasting less than a second. After the mushroom cloud dissipated, radioactive fallout filtered down on Nagasaki (and Hiroshima, destroyed by uranium bomb) and spread on air currents to deliver a second, chronic, lower exposure. People carried, washed in, drank and ate materials, liquids and food that were heavily contaminated by often soluble radioactive elements such as iodine-131, cesium-137 or strontium-90, that once ingested would settle in sensitive areas of their bodies. Japanese medics puzzled over the fact that people who arrived in the bombed cities after the attack grew ill from radiation sickness. American doctors noticed that US soldiers working on reconstruction in Hiroshima and Nagasaki suffered mysterious burns and decreased blood counts.[3] The Japanese press attributed these post-bomb symptoms to a mysterious "atomic poison."

General Groves worried that if the new atomic bomb was banned like chemical and biological weapons, the US government's $20 billion investment would be wasted and the US would be morally condemned [4].[4] Groves issued orders to confiscate Japanese medical records, measurements of radioactivity, notes, slides and films. He sought to frame the atomic bomb as a very powerful conventional explosive and Japanese deaths caused by just thermal burns. Censorship on "residual radiation" (now called fallout) remained in place for decades. "For years," historian Janet Farrell Brodie writes, "radiation remained the least publicized and least understood of the atomic bomb effects" [5].

In 1950, the US Atomic Energy Commission bankrolled and managed the Atomic Bomb Casualty Commission (ABCC), which created the Life Span Study. Investigators eventually included in the study 120,000 Japanese survivors and 75,000 offspring. They tracked vital statistics, cancers and causes of death and coordinated that information against estimates of doses of radioactivity based on subjects' reported location at the time of the bombing, as survivors remembered it. This highly subjective method of retrospective dose reconstruction became the main means to

---

[3] "Memorandum of Telephone Conversation between General Groves and Oak Ridge Hospital, 9:00 a.m., August 25, 1945," National Security Archive, accessed May 1, 2018, https://nsarchive.gwu.edu.

[4] Cost is given in 1996 dollars. See Ref. [4].

judge damage from exposure to nuclear fallout. The first ABCC reports were optimistic. The bomb blasts caused not very alarming increases of a few cancers among survivors.

As early as July 1946, for both technical (miniaturization) and political (impressing Soviets) reasons, the US Armed Forces started testing improved versions of the Pu bombs. In the early 1950s, after the development of the H-bomb, which is ignited by a Pu fission trigger, American and Soviet nuclear testing frequency increased. Americans set up a test site in the Marshall Islands to detonate plutonium-triggered H-bombs. Soviets created a polygon in northern Kazakhstan in Semipalatinsk and in Novaya Zemya. In 1954, several thousand Marshall Islanders were exposed to high levels of radioactivity in the Castle Bravo Test, thousand times more powerful than the Hiroshima bomb. US officials left Marshall Islanders in place for several days before evacuation. Once moved to safer ground, scientists from Brookhaven National Labs gave medical exams but no treatment to the islanders who suffered from nausea, hair loss, fatigue and burns [6]. The Americans were interested in the rates of ingestion and retention in bodies of harmful isotopes. In classified studies, the American scientists recorded thyroid cancers and thyroid disease among 79 % of exposed Marshall Island children under age ten. Anemia in the group was rampant. They also learned that Rongelap women exposed in the Bravo test had twice the number of stillbirths and miscarriages as unexposed women [7]. They did not disclose this information for political reasons.

In 1951, the Atomic Energy Commission opened a new test site in the continental US for fear of supply problems in the Pacific islands due to the Korean War. Between 1951 and 1992, the US government detonated 1,021 bombs, 100 of them into the atmosphere; the USSR detonated 727 bombs, including 219 atmospheric or underwater tests. Public concern about nuclear testing in the US and worldwide grew in the 1950s. While the AEC secretly in "Operation Sunshine" collected human bones and teeth to test for ingestion of radioactive isotopes, the St. Louis' Citizens' Committee for Nuclear Information including Barry Commoner openly collected baby teeth and tested them for radioactive strontium. In the late sixties, University of Utah epidemiologist Joseph Lyon noticed a spike in childhood leukemia and thyroid cancers in areas of Utah downwind of the Nevada Test Site. His report remained buried for several decades. In the 1990s, Lyon finished his study showing elevated thyroid cancers [8].

4. Waste

To produce enough plutonium for a bomb core, manufacturers needed a ton of uranium. After distilling 11 kilograms of plutonium for a bomb, the rest is waste, much of it highly radioactive. At Hanford, minimal resources were initially dedicated to management of growing radioactive waste. The Project had chosen a sparsely populated site. It removed several towns and a Wanapum indigenous settlement from the newly designated "federal reservation" in case of accidents and waste leakage. They coined the platitude: "diffusion is the solution" for nuclear safety. With an elaborate security apparatus, plant engineers drilled holes, called "reverse wells," and dumped in liquid radioactive waste. Workers bulldozed trenches and ponds and poured in radioactive liquid and debris. Cooling ponds held hot and radioactive water for short periods before flushing into the Columbia River. Smokestacks above processing plants issued radioactive particles in gaseous form. High-level waste, terrifically radioactive, went into temporary large, single-walled steel coffins buried underground. The waste would corrode even thick steel walls.

DuPont leaders insisted on some environmental assays to monitor the dumped radioactive waste. At Hanford, they set up a soil study, a meteorological station and a fish lab to monitor the health of salmon migrations in the Columbia River. Meteorologists showed that radioactive gases either traveled long distances in strong winds or hovered over the Columbia Basin caught in an inversion. Soil scientists discovered that plants drank up radioactive isotopes readily from the soil, and aquifers became contaminated. Ichthyologists learned that fish concentrated radioactive waste in their bodies at levels at least 60 to 1,000 times greater than the water in which they swam. Researchers found that radioactive waste did not spread in a diffuse pattern, but collected in random hot spots and readily made their way up the food chain. Despite this troubling news, no major changes were made to waste management practices for 30 years. In the course of 44 years of production, Hanford rolled out the Pu content for 60,000 nuclear weapons [9].

Likewise, Soviet engineers also dumped radioactive waste into the ground, local water sources and the atmosphere. High-level waste was contained temporarily in underground tanks similar to those built at Hanford. Unlike the high-volume Columbia River, the Techa River flowed slowly, flooding frequently and bogging down in swamps and marshes. After two years of Mayak's Plutonium Plant operation, the director called the Techa "exceedingly polluted." By 1949 and the first Soviet nuclear bomb test, the plant's underground containers for high-level waste were overflowing. Rather than shutting down production while new containers were built, plant directors dumped the high-level waste in the turgid Techa River. Four thousand three hundred curies a day flowed down the river. From 1949 to 1951, when dumping ended, 20% of the river consisted of plant effluent, 3.2 million curies of waste clouded the river along which 124,000 people lived. Villagers used the river for drinking, bathing, cooking and watering crops and livestock. Soviet medical personnel investigated the riverside settlements in 1951 and found most every object and body contaminated with Mayak waste. Blood samples showed internal doses of uranium fission products. Villagers complained of pains in joints and bones, digestive track illnesses, strange allergies, weight loss, heart murmurs and increased hypertension. Blood counts were low and immune systems weak. Soviet army soldiers carried out an incomplete evacuation of riverside villages over the next 10 years. Twenty-eight thousand people, however, remained in the largest village, Muslumovo. They became subjects of a large, four-generation medical study. In subsequent years, prison laborers created a network of canals and dams to corral radioactive waste streams into Lake Karachai, now capped with crumbling cement and considered the most radioactive body of water on Earth.

On September 29, 1957, one of the underground waste storage tanks exploded at Mayak. A column of radioactive dust and gas rocketed skyward a half mile. The explosion spread 20 million curies of radioactive fallout over a territory four miles wide and 30 miles long. An estimated 7,500 to 25,000 soldiers, students and workers cleaned up radioactive waste ejected from the tank. Witnesses described hospitals fully occupied and the death of young recruits, but the fate of 92% of the accident clean-up crew is unknown. Following the radioactive trace, soldiers evacuated seven of 87 contaminated villages downwind from the explosion.

Residents of villages that remained were told not to eat their agricultural products or drink well water, an impossible request as there were few alternatives. In 1960, soldiers resettled 23 more villages. In 1958, the depopulated trace became a research station for radio-ecology.

The social consequences of the production and dispersal of large volumes of invisible, insensible radioactive toxins were profound. Plant managers in the US and USSR created special residential communities ("nuclear village" (US) and "closed cities" (USSR)) to manage and control the movement of workers and radioactive isotopes. In exchange for the risks, workers in plutonium plants were paid well and lived like their professional class bosses. They signed security oaths and agreed to surveillance of their personal and medical lives, while they remained silent about accidents, spills and intentional daily dumping of radioactive waste into the environment. Free health care and a show of monitoring employees and sites led workers to believe they were safe. To protect competent employees, much of the dirty work dealing with radioactive waste and accidents fell to temporary, precarious labor (prisoners, soldiers and migrant workers), often ethnic minorities (Muslims in the USSR, Latino and Black in the USA). Segregating space by race to disaggregate "safe" places and sacrifice zones became a larger pattern in Soviet and American Cold War landscapes more generally [10].[5]

In addition to Cold War, waste Pu is also a by-product of nuclear power. All nuclear countries are confronted with the management of an increasing amount of high-level radioactive waste. Natural uranium comprises two isotopes, U-238 (99.3%) and U-235 (0.7%). Uranium-235 nuclei fission after absorbing low-energy ("slow") neutrons: they emit fission fragments and high-energy ("fast") neutrons. In most nuclear reactors, these are slowed down in order to induce further fissions. But when a uranium-238 nucleus captures a high-energy neutron it may, via a fast radioactive sequence, become a Pu-239 nucleus, which undergoes fission efficiently under *fast* neutron irradiation. This is the "breeder" configuration where the entire uranium content is useable for energy production, both directly (U-235) and by transmuting U-238 into Pu-239. Physicists proved that such a system was feasible in 1946 and a number of test versions have operated, but implementation at significant power levels

---

[5] See Ref. [10].

requires an extremely delicate reactor design and extraordinary surveillance to avoid accidentally triggering an explosive chain reaction. Safety, complexity and economics have so far overcome anticipated advantages. So far, the breeder remains a dream.

5. Post Cold War Discoveries

The intense secrecy surrounding plutonium production meant that the public did not focus on plutonium plants as major polluters of radioactive waste. American protestors fixated in the 1950s on fallout from nuclear testing in Nevada and abroad. Later, after the 1979 accident at Three Mile Island, Americans worried about accidents from civilian nuclear reactors, while Europeans protested against the placement of NATO nuclear weapons [11].[6] In the Soviet Union, the word « plutonium » was banned from all records. Soviet citizens did not learn where plutonium was made, nor of location or time of tests until the last years of the USSR.

Interest in Hanford's radioactive legacy emerged in the Reagan years with a renewed nuclear weapons build up. In the early 1980s, when engineers re-opened the mothballed Hanford plant, a whistle-blower contacted a local reporter, Karen Dorn Steele, expressing her fears that the engineers were going to blow the plant up. Steele teamed up with local anti-nuclear activists. The group demanded to see Hanford's environmental and health records. The US Department of Energy finally released 10,000 documents a month before the Chernobyl reactor no. 4 exploded. After Chernobyl, many more American and Soviet citizens got involved in delving into the troubling legacies of plutonium production. Lawsuits mounted. In the United States and USSR, workers received compensation if they had documented exposures and corresponding diseases. People living near the Soviet and American plants, dubbed "Downwinders", did not win lawsuits in court largely because it was impossible to estimate accurately their exposures and link them with cases of cancer and disease. Without numerical evidence of exposure, it was easy in court to deny links between plutonium plant radioactive by-products and disease [12].[7]

---

[6] See, for example, Ref. [11].
[7] On lawsuits and their failure in the US, see Ref. [12].

The Hanford Plant was finally closed in 1989 and in subsequent years it became the United States' largest and most expensive Superfund cleanup site. Reckless waste-storage practices led the US to spend over a 100 billion dollars in 30 years on the cleanup of the Hanford site. The job is not halfway finished. 177 underground tanks contain at least 1,800 contaminants, including plutonium, uranium, cesium, aluminum, iodine and mercury. No two tanks are alike; they have leaked an estimated 4 million liters. Tank cleanup alone will cost a projected 550 billion and take 60 years [13]. Geologic stability and isolation from ground water remain crucial concerns to protect populations from long-term effects (about 300 years for Cs-137 and Sr-90, about 240,000 years for Pu). The estimate for both plants (Mayak and Hanford) is that each plant issued 350 million curies of radioactive waste into the environment, a number that dwarfs emissions from the Chernobyl accident.

These sites were the worst-hit, but by no means the only ones. The Idaho National Laboratory area (larger then the state of Rhode Island) was home to 52 nuclear reactors and other nuclear experimental operations. In the 1950s and 1960s, it became a storage site for waste from these operations. Radioactive waste was buried in unlined pits and trenches or stored in large drums or wooden boxes. The danger here is related to waste seeping into an aquifer feeding water to some 200,000 people. The time for Pu to diffuse into the aquifer was estimated to be 80,000 years in 1965. The solubility of transuranic elements in underground water is often fairly low because the latter's oxygen level is depleted by chemical reactions with surrounding rock. But a more precise evaluation in 1997 revised the time span to only 30 years, clearly indicative of transit paths through faults. Pu, fission fragment radionuclides and other chemicals have recently been detected in the groundwater.

Today, waste-related problems also involve Pu (and other actinides) from civilian nuclear plants. Each plant produces about 20 kilos of Pu per year, most of which is to be potentially "disposed of" by means that are so far neither permanent nor even clearly determined. As noted by Seaborg, Pu's chemistry and physical properties are a major difficulty, requiring a multiplicity of chemical treatments in order to reduce the Pu into a powder form that is often fused into a specially conceived glass form, assumed to be optimally stable, and then buried underground in a "stable" setting. Engineers are dealing with long time spans (hundreds of thousands of years). Stability requires a multi-barrier system in which adequate geological conditions (nature of rock, permeability, potential leaching chemistry, etc.) play a major role [14]. As a result, there is as yet no choice of a final destination for most of American radioactive waste. The high-radiation level Pu and actinides from the military stockpile, on the other hand, are to be stored for 10,000 years in the Waste Isolation Pilot Plant (WIPP), a 650 meter-deep salt dome in New Mexico. So far, the site is about 25% filled; a succession of incidents (leakage of highly radioactive elements, fires due to chemical reactions) has affected the site. Another issue is how to prevent intrusion for a hundred centuries.

In the year 1946, the US produced eight Pu bombs. That number rose to 800-odd in 1952 and 18,638 in 1960. The production rate in the USSR was comparable, with a time delay of about four to five years. Stockpiling a bomb means storing a delicate combination of materials such as high-power chemical explosives, electronic triggering and control mechanisms, plastic isolation and Pu itself, all under slow but permanent irradiation by the radioactive fragments born from Pu fission. Safe stockpiling of the primitive weapons was a major problem until the 1970s. Recent decisions regarding "modernized" (i.e., smaller, computer-brained) weapons will revive the dangerous Pu arms fabrication industry.

6. Conclusion: Plutonium's Impact

Plutonium profoundly changed the world. Military leaders created a weapon with a toxic legacy that defies human comprehension. People learned to fear imminent nuclear apocalypse and a slow death from invisible contaminants. Anxious about rogue attacks, officials created elaborate security apparatus — fences, surveillance, intelligence, criminalization of everyday acts, spatial segregation. These security features expanded across the landscape so that today entry into an airport shares features that early Hanford workers underwent to gain access to the plant. Most people have become used to surveillance through social media of their movements and private correspondence. Plutonium and its radioactive by-products are so widespread that most humans in the global north carry traces of the nuclear arms race in their bodies. Humans in the 21st century can no longer separate their lives, cultures and societies from that of plutonium.

References

1. Brown, K. (2013). *Plutopia: Nuclear Families, Atomic Cities, and the Great Soviet and American Plutonium Disasters.* New York: Oxford University Press, p. 65.
2. Groueff, S. (1967). *Manhattan Project.* New York: Little & Brown.
3. Brown. *Plutopia.* pp. 174–175, 178–180.
4. Schwartz, S. I. (1998). *Atomic Audit: The Costs and Consequences of U.S. Nuclear Weapons since 1940.* Washington, DC: Brookings.
5. Brodie, J. F. (Summer 2015). Radiation secrecy and censorship after Hiroshima and Nagasaki. *Journal of Social History.* 48(4), 842–964.
6. Johnston, B. R. and Barker, H. M. (2008). *Consequential Damages of Nuclear War: The Rongelap Report.* Walnut Creek, CA: Left Coast Press, pp. 173–191.
7. Smith-Norris. (2016). *Domination and Resistance: The United States and the Marshall Islands During the Cold War.* Honolulu: University of Hawaii Press, pp. 89–93.
8. Ball, H. (1986). *Justice Downwind: America's Atomic Testing Program in the 1950s.* New York: Oxford University Press. pp. 158–72; Leopold, E. (2009). *Under the Radar: Cancer and the Cold War.* New Brunswick, NJ: Rutgers University Press, pp. 169–80.
9. Gallucci, M. (April 28, 2020). A glass nightmare: Cleaning up the cold war's nuclear legacy at Hanford. *IEEE Spectrum.*
10. Hales, P. B. (1997). *Atomic Spaces: Living on the Manhattan Project.* Urbana: University of Illinois Press; Farish, M. (2010). *The Contours of America's Cold War.* Minneapolis: University of Minnesota Press.
11. Tompkins, A. S. (2016). *Better Active than Radioactive! Anti-Nuclear Protest in 1970s France and West Germany.* Oxford: Oxford University Press.
12. Pritikin, T. T. (2020). *The Hanford Plaintiffs Voices from the Fight for Atomic Justice.* Lawrence, Kansas: University Press of Kansas.
13. Shoop, D. S. (January 31, 2019). Completion of the 2019 Hanford lifecycle scope. (Accessed 2020 June 9). https://www.hanford.gov/files.cfm/2019_Hanford_Lifecycle_Report_w-Transmittal_Letter.pdf.
14. Lalenti, V. (2014). Adjudicating deep time: Revisiting the United States' high-level nuclear waste repository project at Yucca mountain. *Science & Technology Studies.* 27(2), pp. 1–22

Part 4

Critical Materials

Bernadette Bensaude-Vincent

Materials become critical when there is a risk of supply disruption that will have a tremendous impact on geopolitics and global economy. Those materials are considered strategic because the country holding these vital resources may acquire power over others or be doomed because it attracts imperialist powers. The life story of rubber illustrates the close connection between the desire for seizing materials and colonial empires. It also demonstrates that even renewable raw materials — like plants — can be critical in specific circumstances, like war or economic blockade.

Most often a material becomes critical because of its association with a dominant technology. Rare earths are critical because they are used in electronic devices and solar panels. Lithium is critical today because it is used in batteries, which are more and more in demand for energy storage in electric vehicles or intermittent sources of power like windmills. So criticality is tightly related to technological choices.

The criticality of materials is not just a matter of physical scarcity at the beginning of the supply chain. The risk of geological depletion of a mineral that is a vital component in a successful technology — like rubber for tires in automotive industry or copper for electric and electronic industries — is only one facet of the problem. Critical is the raw material whose extraction and processing cause health and environmental damages. The environmental damage can be indirect: the extraction and treatment of lithium, for instance, consumes a large part of the water resources of Chili, thus causing serious restrictions for the population.

In addition to the geological and environmental dimensions, the criticality of materials has also a geopolitical dimension when one country, or a few companies, hold most of the global resources. For instance, half of the extraction and refinement of lithium is in the hands of three transnational companies. In the case of rare earths, their criticality is not so much a question of rarity; it is rather that one country — China — holds 60% of the global resources.

As a general rule, when a material appears as *matter of interest* for an emerging technological sector, it soon becomes a *matter of affairs* for a number of investors, states or industrial corporations. And it ends up as a *matter of concern* because of its impacts on populations, on international relations and on the environment. Everyone who has seen the mining landscapes where metallic ores locked for millennia within rock have been extracted for a few decades, or has experienced the work and health conditions of the population of miners, is aware of the human and environmental costs of the comfort provided by our technological devices.

More sustainable technologies should reduce the list of critical materials. Recycling is the most obvious way to reduce the criticality of materials. However, the processes of disassembling and isolating the individual components of a chip or an electric vehicle battery produce waste, often use hazardous solvents, and emit greenhouse gases. To ease the process of recycling, manufacturers have to start designing their products with recycling in mind. Critical materials thinking thus requires more than just technological advances. Logistic solutions, such as placing the recycling centers in the right place, are needed as well as political measures and strong regulations to compensate the high cost of recycling.

Chapter 13

Rubber

Stephen L. Harp

University of Akron (Ohio), USA

Rubber is, above all, flexible, and its past has proven malleable in human societies. After taking the form of an elastic ball for ritual games in the Western hemisphere, rubber became a strategic global material in the 20th century — before largely disappearing from public view as a commodity in the 21st century.

1. Bouncing Balls

For centuries, Mesoamerican societies had harvested latex from what later became known as the *castilla elastica*, a tree found in what is today southern Mexico and Central America. Latex's most important use, as described by the Spanish, was in fashioning balls for ritual games. At Moctezuma II's court in Tenochtitlan (today's Mexico City), Spanish conquistadors watched the complex game, especially marveling at the bouncing ball, so much so that they hauled both players and balls back to Spain to serve as live exhibits for the Spanish court.

Inhabitants of the Amazon river basin learned independently of Central American peoples how to tap what Europeans later dubbed rubber trees, including the *hevea brasiliensis,* or simply *hevea.* Tupi-speaking Indians in what is now Brazil called that tree *cahuchu,* literally "wood that weeps," variants of which became the word for "rubber" in several European languages: *caucho* in Spanish, *caoutchouc* in French and *Kautschuk* in German. Amazonians fashioned coagulated latex into a series of products, notably boots, which were obviously very useful in the tropical rain forest. Little by little, over the next two centuries, Europeans on scientific expeditions learned more about the mysterious substance, seeing the actual latex-bearing trees, how they were tapped, and how Indians transformed the latex into objects. On an expedition in South America, Frenchman Charles Marie de la Condamine saw a rubber tree tapped and named the whitish sap-like substance "latex" (Latin for liquid or liqueur) and the smoke-cured result *caoutchouc*. In the late 18th century, British natural philosopher Joseph Priestley (also known for first isolating oxygen) used a nub of the substance to erase, or "rub out" pencil marks, and named it "India Rubber," a designation that eventually became simply "rubber" in English.

2. The Substance of Colonial Empires

In the 19th century, rubber, a product of empire, became a product of industry. In the 1820s, the British chemist Charles Macintosh used coal tar naphtha to dissolve solid rubber (and then apply it to canvas to make raincoats, henceforth called "Macintoshes"). When combined with Thomas Hancock's patented "masticator," which could chew up solid balls of cured rubber shipped from the South America, the manufacture of rubber products became possible outside rubber-growing regions. Up to this point, the latex had generally been smoke-cured into objects, such as boots or balls, on site in South America, then shipped to Europe or North America. Now, however, solid balls of smoked rubber could be broken down, easily dissolved, and made into objects in the northern hemisphere.

By the late 1830s, American Charles Goodyear found that the addition of sulfur to heated masticated rubber would keep the resultant rubber products from melting in the heat and cracking in the cold. Goodyear got a U.S. patent for the process he had discovered, which he named vulcanization after the Roman god of fire. In 1851 in London, at the first world's fair, the Crystal Palace Exhibition, rubber manufacturers exhibited a host of rubber goods in an effort to build a market. There Thomas Hancock and other British manufacturers showed toys, Macintosh cloaks, capes, pillows, cushions, life preservers, model pontoons and assorted other rubber products. Goodyear set up a large stand with myriad articles called "Goodyear's Vulcanite Court." Goodyear displayed walls, furniture, jewelry, household goods and medical instruments of ebonite (hardened rubber later used for telephone casing and other products before the development of modern plastics).

Less noticed at the time were the industrial uses of rubber. In the steam engines powering the factories that built weapons, the steamships that carried European troops and indigenous laborers back and forth across empires, and the railway engines that moved men, women and material in Europe and increasingly in European colonies, rubber was the raw material for a host of industrial parts. Washers, gaskets, buffer and bearing springs, rolling pistons, plug valves, hoses, belts, motor mounts and other unseen rubber parts became key components of both advanced industry and the imperialism that it made possible.

The 1880s and 1890s witnessed two developments that were inextricably tied. First, industrialization intensified after the start of the Second Industrial Revolution, during which ever more manufactured goods found their way to ever larger numbers of Americans and Europeans who could afford to purchase them. Two key consumer products born of the Second Industrial Revolution were bicycles and automobiles. Second, these same years saw the heyday of European empires in Africa and Asia.

3. Huge Plantations and Exploitative Work

While most of the world's rubber came from Brazil, the African rubber trade grew dramatically in the 1890s, with European imperial powers attempting to profit from it. On that continent, native species of latex-producing plants included the tree *funtumia elastica* (sometimes known as Lagos silk rubber) and the woody *landolphia* vines that grew into the branches of other trees. In some areas, such as the Kongo area of what is today Angola, Africans remained almost entirely in control of trade in rubber [1]. More notoriously, in the Congo River Basin, particularly in Belgian King Leopold II's Congo Free State, state-sponsored companies received land as concessions for exploitation, where they forced Africans to gather rubber. Leopold's men oversaw an armed force of African troops, called the *Force publique,* to enforce rubber collection; poorly paid and brutalized by white superiors, they in turn treated local Congolese horribly. Although we have no census data for 19th-century Africa, estimates of deaths of indigenous peoples gathering rubber in the Congo Free State range as high as 10 million [2].

In the 19th century, colonial powers honed systems to encourage the economic exploitation of their empires. Agricultural development was central, and plants moved to and fro across the globe. Among the many efforts to gain control over global plant life, the British Royal Botanic Gardens at Kew (the largest in the world and a model for botanical gardens everywhere) contracted with several Britons with contacts in Brazil to gather hevea seed. In 1876, Henry Wickham took a shipment of some 70,000 hevea seeds from Brazil to Kew. Kew was literally a hothouse from which plants were disseminated to other, smaller botanical gardens of the empire.

In a sense, Kew was the hub, and the smaller gardens the "spokes" of an elaborate system of government-subsidized agricultural science and production [3]. Eventually, seeds from Kew resulted in seedlings sent to Asia, some of which ended up at the Singapore Botanical Gardens.

It was in Singapore that botanist Henry Nicholas Ridley studied and propagated the hevea trees. In Brazil, each hevea tree grew in isolation, only about two per acre, a naturally occurring spacing that prevented the tree-to-tree transmission of a microscopic fungus widespread in Brazil. Ridley found, however, that the hevea trees could grow on plantations in Southeast Asia, where the fungus was not a major threat. Ridley led an effort to develop a science of the rubber tree, asserting that the hevea was the most economically viable source for the latex needed to make the high quality rubber increasingly demanded by industrialists. By 1899, more than a million seedlings had been planted on the Malay Peninsula.

Hevea trees proved an ideal plantation crop in Southeast Asia. Requiring tropical heat and about seventy inches of rain per year in order to thrive, the trees seemed perfect for the huge tracts of rainforest that colonial governments had been granting planters as virtually free concessions, with long-term leases for coffee, tapioca, sugar, pepper, tea, tobacco and now rubber cultivation. Colonial governments also offered planters subsidized loans. In European eyes, the "jungle" was essentially vacant land, even in areas where indigenous peoples had long been using it for swidden agriculture.

Southeast Asian plantations enjoyed other advantages. Nearby shipping lanes facilitated transportation of rubber back to Europe and directly to North America. Nearby were sources of what advocates invariably called "cheap, abundant labor," from China, the Indian Subcontinent and Java. By the turn of the century, British planters had imposed rubber plantations on the Malay Peninsula in what is today Malaysia, using Chinese and Tamil laborers. The Netherlands East Indies encouraged international investment, so British, American and other nationalities joined the Dutch in founding plantations in what is today Indonesia, particularly on the island of Sumatra; here most laborers were Javanese and Chinese. The French followed suit in Indochina, creating plantations in southern Vietnam and Cambodia, employing laborers shipped down from the densely populated northern Vietnamese province of Tonkin.

Plantation rubber all but replaced collected or "wild" rubber in world markets. In 1900, of the 59,326 tons of rubber produced in the world only 3 tons originated in Asia. In 1919, of the 465,845 tons produced in the world, 420,046 came from Asia, almost all from rubber plantations [4]. Given the costs of clearing, planting and then awaiting the first harvest, rubber plantations were capital intensive, and big firms signed leases on huge tracts of land. Some of the largest rubber plantations were owned by tire companies, which wanted to ensure a steady supply of affordable rubber, whatever the price fluctuations for the commodity in global markets.

Plantations became huge and grew larger. In 1910, Dunlop acquired 50,000 acres in Malaya and added another 10,000 or so by 1917. Even after World War II, Dunlop's Malaya plantations remained the single largest private landholding in the entire British Commonwealth [5]. American firms preferred Sumatra because the Dutch encouraged international investment in order to use foreign capital to "develop" the island. In 1910, the American rubber and tire conglomerate U.S. Rubber took over the leases on 88,000 acres on Sumatra and controlled more than 110,000 acres (150 square miles) in Sumatra and Malaya in 1926, and 135,000 acres in 1937 [6]. In 1917, Goodyear had 16,700 acres, and 54,700 in 1932 [7]. By

1927, the French-owned Michelin Tire Company's plantations of Dâu Tiêng and Phú Riêng in southern Vietnam consisted of some 21,750 acres and 13,750 acres, respectively (of about 417,500 acres of rubber plantations in French Indochina generally) and employed more than 4,000 laborers [8].

Working conditions were harsh. Particularly in the early days of the concerns, during the clearing of the land for planting, laborers' death rates were shockingly high. In 1900, in Sumatra, mortality rates briefly ranged as high as 23.8% per annum [9]. In 1926, the mortality rate on the Société Indochinoise des Cultures Tropicales' Budop plantation was 47% per annum [10].

4. A Patriotic Material

In the cataclysm that was the Great War, rubber proved its mettle. In September 1914, French forces halted the German onslaught, saving Paris from invasion. Trucks and government-requisitioned Renault taxicabs from Paris hauled men and materials to the front in the famous French victory at the Battle of the Marne. In 1916, Verdun was the site of a long, brutal battle that came to symbolize the entire French war effort (about 80% of French troops served for a time at Verdun, 300,000 of whom died). There the "sacred way" (*voie sacrée*) received much credit for victory. On that narrow road, trucks ran round the clock, bumper to bumper, in an endless convoy to and from the front, carrying men and supplies. After the battle of Verdun, the future seemed clear. Tire-clad trucks would be critical in future wars. Meanwhile the Germans, cut off by the British blockade from rubber supplies, faced a severe shortage of the precious commodity, prompting German chemists to begin experimentation in the production of synthetic rubber, with very mixed results. The lack of rubber did not cause German defeat even if abundant rubber assisted in French and British victory. But the lesson had not been lost on the leaders of the major combatant powers. Future wars would be fought largely on rubber.

After World War I, the supply of rubber remained the object of nationalist arguments, particularly between Britain and the United States. The war had spiked demand for various raw materials, including rubber.

During the course of the war, planters in Southeast Asia planted ever more hevea trees. After the end of the conflict the tables turned, and, with demand unable to keep up with an increasing supply, rubber prices plummeted. In 1922, rubber fell below 12.5 cents per pound, below the cost of production. Rather suddenly, many planters faced bankruptcy. In 1920s, British-controlled Malaysia and Ceylon produced over 70% of the world's rubber supply while the United States consumed more than 70% [11].[1] For the British Rubber Growers' Association, a powerful planters' lobby, the solution seemed simple. Britain, the self-proclaimed prewar champion of free trade, needed to protect British planters by increasing the price of rubber at the expense of manufacturers. Secretary of State for the Colonies Winston Churchill agreed and established a committee headed by Sir James Stevenson, a Churchill crony who had made his name and much of his fortune as managing director of the Johnnie Walker whisky company.

The Stevenson Committee considered various means for raising the price of rubber and proposed legislation establishing rubber production quotas. Parliamentary passage was swift, as nationalist sentiment was strong. Britain had borrowed heavily from the United States to fight World War I and British members of parliament thought it only fair that "British rubber" (from British-controlled colonies) fetch a decent price. Churchill specifically cited British war debts in defending the plan. As a result of the legislation, all rubber plantations in British colonies, even those owned by foreign companies, received precise quotas for rubber production.

Within the British Empire, participation in the Stevenson plan was obligatory. In Malaysia, Chinese and Malay smallholders who had recently planted rubber groves were forbidden to tap or expand production, a ban that led to both resistance and smuggling. Planters in the Netherlands East Indies and French Indochina opted not to participate. As rubber prices rose as a result of the Stevenson scheme, planters in both of those colonies expanded their acreage. In French Indochina, Michelin and investment companies leased and planted vast new tracts of land. In the Netherlands East Indies, indigenous smallholders significantly extended

---

[1]Memorandum Submitted by the Rubber Manufacturers' Association of America to the Department of Commerce, 16 July 1925, in Commerce Papers: Rubber, Box 531, Herbert Hoover Presidential Library; See Ref. [11]

the size of their rubber groves, tapped them and profited handsomely. In the short term, however, the policy worked for British growers. In 1925, rubber was selling at $1.23 per pound [12].

Although British nationalism, as stoked by planters, was critical in passing the Stevenson restrictions, U.S. responses were more fiercely nationalistic. Herbert Hoover and rubber magnate Harvey Firestone led the charge. Hoover served as the Secretary of Commerce from 1921 to 1928 under Presidents Harding and Coolidge. Under his leadership, the U.S. Commerce Department forcefully asserted itself in the protection of U.S. business interests abroad. That began with ensuring the supply of raw materials for industry. In 1921, the Commerce Department created an entire division devoted exclusively to rubber. Secretary Hoover personally took a major role, meeting frequently with the American Rubber Manufacturers' Association (founded in 1915 and dominated by major U.S. tire firms) and individual company leaders, particularly Harvey Firestone. A darling of the press for his earlier work in war relief, Hoover brilliantly courted journalists. Like Firestone, Hoover referred frequently to "our rubber" and argued that "America needs its own rubber." He thus cast himself as a champion of "free enterprise" in attacking "foreign monopolies."

Hoover attacked the Stevenson plan on several fronts. Much as Americans had been exhorted to "Hooverize," or conserve, during World War I, now they were to exercise their patriotism by turning in old rubber products for reprocessing. Moreover, after meeting with Firestone, Hoover strongly supported the latter's plan for a Congressional appropriation of $500,000 largely to undertake a series of "rubber surveys" in different parts of the world. These studies focused on the Philippines, Central and South America, Africa and Southeast Asia, laying out the various conditions that would help or hinder the establishment of American-owned plantations in each place. Ultimately, the U.S. government encouraged Firestone's ambitions in Liberia. Liberia had been a virtual colony of the U.S. since the 19th century. Here the climate was appropriate, labor was cheap and the U.S. government would be able to protect investments. Firestone acquired a lease there for one million acres.

After peaking at well over $1, the price of rubber hovered at 50 cents a pound in 1926. Hoover publicly declared victory, taking credit for the decline. As rubber prices continued to fall, a fact mostly attributable to increased production on the part of indigenous smallholders in the Netherlands East Indies (a 900% increase between 1920 and 1930), the British government realized the limits of the Stevenson plan and repealed it in 1928. Having wrapped himself in the American flag, Herbert Hoover was elected president in November 1928.

Firestone was a close friend of Henry Ford and Thomas Edison. The three men talked extensively about the rubber crisis that the United States supposedly faced due to the Stevenson plan. Personally taking the charge to "grow America's own rubber," Edison devoted the last years of his life to botanical expeditions and experiments in an effort to find an adequate latex-yielding plant well suited to growing conditions in the United States. In 1927, Firestone, Ford and Edison each contributed $25,000 to establish the Edison Botanic Research Corporation, a botanical garden and rubber laboratory at Edison's winter home in Ft. Myers, Florida. Edison and his staff ultimately chose goldenrod over the slower-growing guayule as the best alternative source, but neither plant ever proved commercially competitive on a scale that could rival the hevea tree.

For his part, Ford bought just less than 2.5 million acres in the Tapajós valley in Brazil and set up a plantation modestly named after himself, "Fordlandia," and a second named Belterra. Creating a veritable American town within the rainforest, Ford attempted to grow hevea trees to supply rubber for his factories. The result was a complete fiasco. Leaf-blight and pests ruined the trees. Ford's treatment of Brazilian laborers, including a classically paternalistic Fordian ban on alcohol and tobacco use even in laborers' dwellings, led to significant conflict [13].

5. The Quest for Synthetic Rubber in World War II

During World War I, the British naval blockade effectively halted rubber from making its way into Germany. The wartime German rubber shortage was acute, despite two famously daring trips of the submarine *Deutschland* through the blockade to the United States (still neutral before April 1917, its ports in Baltimore, Maryland and New London, Connecticut welcomed the sub), where they delivered dyestuffs and pharmaceuticals in return for rubber. Rubber sold for the equivalent of $20 per pound in Germany, compared to about $1 per pound in the United States. German military hospitals soon ran out of rubber gloves, tubes and syringes, and there was a severe shortage of truck tires.

Germany was the world's chemical powerhouse, providing most of the textile dyes, artificial fertilizer and pharmaceuticals (such as Bayer aspirin) consumed in the United States and elsewhere. German chemists worked to come up with substitute ("ersatz") rubber, just as they had developed ersatz versions of other products. They only managed to develop a form of synthetic rubber that was essentially useless, too hard even for use as truck tires. Still, research continued in the 1920s at IG Farben, the huge chemical conglomerate that brought together Bayer and so many other firms in 1925. In the late 1920s, IG Farben chemists worked with butadiene, combining it with styrene to create Buna, a more pliant synthetic rubber.

When Hitler came to power in 1933, he immediately began to plan for war. To avoid a repetition of World War I, he demanded autarky, especially in metals and rubber. He turned to IG Farben, the largest company in Europe, to refine and develop Buna for military use. Germans used two raw materials that they had in abundance, coal and limestone, to create acetylene base for Buna. Chemists at IG Farben shortened the production time and found that though still less elastic than natural rubber, synthetic rubber was actually better at resisting the deleterious effects of sunlight and petroleum products including gasoline, a decided advantage on the battlefield. The problem was cost. German economists estimated that a tire made of Buna would cost 90 marks to produce, compared to 18 marks for one made of natural rubber [14]. Clearly, a five-fold price difference was not justifiable, unless the purpose of developing Buna was for war. Hitler personally authorized a significant subsidy to bring Buna tires to market.

Buna made World War II possible. Over the course of the war, IG Farben continued to refine synthetic rubber and mostly replaced Germany's use of natural rubber. The USSR continued to ship natural rubber to Germany before the German invasion in 1941, but after that date the Nazi war machine had virtually no natural rubber beyond scrap rubber. "In late 1941, German tires contained an average of 37% natural rubber, but this declined to around 8% in February 1943 [15]."

The United States had not developed and marketed synthetic rubber. Hevea rubber was extremely plentiful and cheap in the 1930s. In the last years of that decade, the United States began to stockpile rubber, as it seemed critical for a war effort that was beginning to seem inevitable. Stockpiling intensified in 1940 and 1941. In early 1942, just after the attack on Pearl Harbor, the Japanese invaded and occupied British and Dutch colonies. The Japanese cut the United States off from about 97% of its prewar supply of rubber.

In the United States, which had stockpiled proportionately less rubber than had Britain, the result was panic. The U.S. simply could not fight this war it had entered without rubber:

Each Sherman tank — and the United States eventually produced 50,000 of them — required about a half ton of rubber. Each of the nation's heavy bombers needed about a ton. Each battleship contained more than 20,000 rubber parts, totaling about 160,000 pounds on each ship. Americans produced 1.4 million rubber airplane tires in 1944 alone. American soldiers wore 45 million pairs of rubber boots, 77 million pairs of shoes with rubber soles, and 104 million pairs of shoes with rubber heels. Every industrial facility contained rubber conveyor belts and wheels [16], not to mention rubber gaskets within the machines themselves, let alone the tires for farmers and truckers on the home front.

Anguished articles appeared in the press. President Roosevelt asked citizens to gather their scrap rubber and turn it into their local gas stations for a penny a pound. Roosevelt also imposed a national speed limit of 35 miles per hour to conserve tires. Conservation was the first and obvious step, though it was wholly inadequate. As the Rubber Survey Committee, created by Roosevelt to address the rubber crisis, succinctly put it: "Of all the critical and strategic materials, rubber is the one which presents the greatest threat to the safety of our Nation and the success of the Allied cause … if we fail to secure quickly a large new rubber supply, our war effort and our domestic economy will collapse [17]."

The solution was synthetic rubber. Suspending the enforcement of anti-trust law, the federal government worked with rubber companies and invested heavily in the development and production of synthetic rubber. One estimate had the total governmental investment at some $750 million.[2] Initially, grain alcohol served as the base for American synthetic rubber, as this made the process simpler and the Midwest had a grain surplus. By the end of the war, petroleum served as the base for synthetic rubber, as it would in the postwar years. The manufacturing process for using petroleum was more complex, but petroleum was cheaper. Ultimately, U.S. firms built fifty-one synthetic rubber factories between 1942 and 1945. During the same period, production climbed from 24,640 tons of synthetic rubber in 1942 to more than 784,000 tons in 1945.

The rapid development of synthetic rubber in the U.S. was astonishing. In retrospect, the gap between the U.S. and Germany, where Buna production merely doubled between 1941 and 1944, is remarkable. Why the difference? When the head of IG Farben, Carl Bosch, told Hitler in 1932 that the dismissal of Jewish scientists from IG Farben "could set back German science by a century, Hitler replied, 'Then we'll work a hundred years without physics and chemistry [18].'" All Jewish scientists had been forced out of IG Farben by 1937. Scientific research depends on talent, and many of the most talented engineers and scientists left Germany in the 1930s.

When we move from research to production of synthetic rubber, the gap between U.S. and German experiences becomes even clearer. German wartime rubber production suffered from the regime's imperial racial ambitions. In 1941, IG Farben began construction of a huge Buna plant, projected to supply all of Germany's needs. Near Auschwitz, in occupied Poland, the company found the necessary coal and limestone. The site was at the time outside the range of Allied bombers. There was abundant water for processing. And there was abundant cheap labor, as IG Farben contracted with the SS for slave labor. Near the town of Monowitz, this new Auschwitz camp, about five miles from the gas chambers, was often referred to simply as "Buna." As many as 20,000 inmates were at work at Monowitz at any given time, and as many as 300,000 worked there during the war. The horrific conditions at Monowitz have been immortalized by Primo Levi, who described his experiences there in his well-known

---

[2]Report on synthetic rubber, 1946, in Goodrich files, JBL-35 in the University of Akron Archives.

Survival in Auschwitz. At Buna, laborers were quite literally worked to death at the construction site.

Yet, in the words of Primo Levi, "the Buna factory, on which the Germans were busy for four years and for which countless of us suffered and died, never produced a pound of synthetic rubber [19]." Significantly, while decently paid American workers rapidly built synthetic rubber facilities, at Buna most workers were slave laborers. Literally starved to death (average weight loss was 6.5-9 pounds each week), they were not particularly productive. In 1941, as building began, IG Farben carefully calculated that concentration camp inmates would only be about 75% as productive as well-fed workers, though in reality 33% seems to have been more accurate [20]. Survivors reported efforts to sabotage the building site. By 1944, Allied bombers also took a toll. The contrast with the U.S. experience was marked.

6. Conclusion

Though still a strategic commodity, synthetic rubber precluded rubber shortages in the post-World War II world. When oil prices rise, synthetic rubber becomes more expensive, and demand for natural rubber increases. When oil prices fall, producers generally use more synthetic rubber, while demand for natural rubber falls. As a result, in contrast with the first half of the 20th century, the history of rubber is no longer visible in the media [21]. Instead, rubber is now a commodity that is largely behind the scenes, in contexts where we take it for granted. Rubber is still present in industry, medicine, transportation and the bedroom, in various forms. But rubber is no longer in the strategic spotlight.

References

1. Vos, J. (2011). Of stocks and barter: John Holt and the Kongo rubber trade, 1906–1910. *African Slavery in Latin America and the Caribbean*, 19, 153–175.
2. Hochschild, A. (1998). *King Leopold's ghost: A story of greed, terror, and heroism in colonial Africa*. Boston: Houghton Mifflin; Coquéry-Vidrovitch, C. (1972).

Le Congo au temps des grandes compagnies concessionnaires, 1898–1930, Paris: Mouton; Vanthemsche, G. (2012). *Belgium and the Congo, 1885–1980*. Trans. Alice Cameron and Stephen Windross. Cambridge: Cambridge University Press, pp. 23–24.

3. Musgrave, T. and Musgrave, W. (2000). *An Empire of Plants: People and Plants that Changed the World*. London: Cassell, p. 149.

4. Santos, R. (1980). *História econômica da Amazônia, 1800–1920*. Sâo Paulo: T.A. Queiroz, p. 234.

5. McMillan, J. (1989). *The Dunlop Story: The Life, Death and Re-Birth of a Multinational*. London: Weidenfeld and Nicolson, p. 147.

6. Yacob, S. (2007). Model of welfare capitalism? The United States Rubber Company in Southeast Asia, 1910–1942. *Enterprise and Society*. 8(1), 143, 145; Babcock, G. D. (1966). *A History of the United States Rubber Company: A Case Study in Corporation Management*. Bloomington, IN.: Graduate School of Business, p. 87.

7. Stoler, A. L. (1985). *Capitalism and Confrontation in Sumatra's Plantation Belt, 1870–1979*. New Haven, CT: Yale University Press, pp. 18–19.

8. Annuaire du Syndicat des Planteurs de Caoutchouc de l'Indochine. (1926). Saigon. pp. 83–84; Annuaire du Syndicat des Planteurs de Caoutchouc de l'Indochine. (1931). Saigon, pp. 17–33, 55, 85.

9. Stoler. (1985) [note 7], p. 34.

10. Panthou, E. (2013). *Les plantations Michelin au Viêt-Nam: Le particularisme des plantations Michelin*. Clermont-Ferrand: Editions La Galipote, p. 244.

11. Holt, E. G. (1928). Chief rubber division, department of commerce. Five years of restriction. *India Rubber World*. 77(5), 55–57.

12. Ibid., pp. 55–57.

13. Grandin, G. (2009). *Fordlandia: The Rise and Fall of Henry Ford's Forgotten Jungle City*. New York: Metropolitan Books/Henry Holt, p. 350.

14. Jeffreys, D. (2008). *Hell's Cartel: IG Farben and the Making of Hitler's War Machine*. New York: Metropolitan Books/Henry Holt.

15. Tully, J. (2011). *The Devil's Milk: A Social History of Rubber*. New York: Monthly Review Press, pp. 293–295.

16. Finlay, M. R. (2009). *Growing American Rubber: Strategic Plants and the Politics of National Security*. New Brunswick, NJ: Rutgers University Press, p. 171.

17. Chalk, F. R. (1970). The United States and the international struggle for rubber, 1914–1941. Ph.D. dissertation, University of Wisconsin, p. 237.

18. Borkin, J. (1978). *The Crime and Punishment of IG Farben*. New York: The Free Press, p. 57.

19. Levi, P. (1996 [1958]). *Survival in Auschwitz: The Nazi Assault on Humanity*. Trans. Woolf, S. New York: Simon and Schuster. p. 73.

20. Borkin, *The Crime and Punishment*, pp. 117–127.

21. Harp, S. L. (2016). *A World History of Rubber: Empire, Industry, and the Everyday*. Chichester: Wiley Blackwell, pp. 137–141.

Chapter 14

Lithium

Pierre Teissier

Nantes Université, Centre François Viète, France

1. Introduction

Lithium belongs to the short list of elements synthesized during the first three minutes of the universe. According to Big Bang models, the "gradual cooling of matter led to [... a] well-known mixture of primordial gas with a mass proportion of hydrogen to helium (H:He) of approximately 0.75:0.25 and a tiny admixture of lithium and an even tinier of heavier elements [beryllium]" [28, p. 172]. One hundred million years later, it was produced by stellar nucleo-syntheses, scattered throughout galaxies, and incorporated into planetary bodies. On Earth, it participated in the evolutionary process of life by bridging gaps between inorganic and organic domains through "multiple physiological manifestations": ionic transport, modulation of immune systems, stimulation of tissues' growth, effects on metabolism and defenses against viruses and dermatological disorders [3, pp. vi–viii]. All these skills remained unnoticed until late in human history.

When it entered the chemical stage in the late 1810s, lithium was just a walk-on among dozens of simple bodies identified by analytical chemists. Indeed, if Antoine-Laurent Lavoisier listed 33 simple bodies in 1789, seven decades later, Dmitri Mendeleev classified around 70 elements in his famous 1869 periodic table. In this chemical race led by Humphry Davy in London and Jacob Berzelius in Stockholm, lithium was found in the private laboratory of the latter by Johan August Arfwedson. This "mineral alkaline" was isolated while analyzing silicate compounds from the iron mine of Utö near Stockholm [7, p. 45]. The name "lithion" (later lithium), which derived from the Ancient Greek word *lithos*, meaning stone, was forged to refer to "its first discovery in mineral kingdom". This generic etymology echoed the remarkable discretion of lithium until then. Lithium was inscribed in the electro-chemical classification of Berzelius: it was trapped with 30 other bodies in the class of "electro-positive metals" by opposition to (12) "metalloids" and (9) "electro-negative metals". This soft silvery metal was known to combine with electro-negative bodies (halogens, oxides or sulfurs) to give various solid salts with a high fusibility.

This dull existence lying down in textbooks changed dramatically when lithium met two fields of application in psychiatric treatments and energy storage during the 20th century. The chemical element was thus metamorphosed into two materials: one, a psychotropic drug for medicine and the other, as battery storage for industry. Its material life, which used to be framed by measurement and purification processes, was recomposed to provide broad and deep actions on nervous and electric systems. The lithium-psychotropic drug was extensively used to regulate bipolar disorders of human patients in spite of a number of side effects on brains and bodies while the lithium-ion battery was extensively used to flexibilize the system of electronic devices in spite of undesirable side effects of lithium's massive extraction on sub-soils and local communities.

This biography intertwines the two social roles taken by lithium in medicine and industry. The following part compares the two historical narratives of lithium, as psychotropic and energy materials, that carry modern myths. The two lithium-based materials are then portrayed at an epistemic level through their ability for crossing the boundaries between nature and culture, between biological, organic, and mineral domains. The third part analyses the global supply chain of lithium for commercial products. This leads to characterize, finally, the two criticality of lithium:

in sociopolitical terms for energy storage; in ethical terms for psychotropic treatments.

1. The Two Lives of Lithium

Historical narratives on psychiatric and energy materials were written and controlled by practitioners rather than historians. This is typical of academic communities, to build their professional identity upon a collective memory [1]. On the psychiatric side, an "official version" of the history of lithium is "familiar to almost everyone who prescribes or carries out research on lithium" [27, p. xvii]. *The History of Lithium Therapy* by Neil Johnson, an English physician, distinguishes three periods: (1) the first uses of chemicals in psychiatry (1850–1900), including the use of lithium bromides to treat mood swings (1870–1900); (2) the "crossing of the desert" for lithium as a psychotropic drug (1900–1950); (3) the successful spread of lithium to treat bipolar troubles (also known as manic-depressive disorders) following the "pioneer" works of the Australian psychiatrist John Cade, in 1949, continued in the 1950s by two Danish psychiatrists, Poul C. Baastrup and Mogens Schou. The historical narratives by practitioners highlight heroic figures (Cade, Schou, Baastrup), moral behaviors (stubbornness) and good practices (interdisciplinary and international approach). A recent book by a former professor of Brown University, Walter Brown, titled *Lithium: A Doctor, a Drug, and a Breakthrough*, tells a "success story" that features John Cade as a hero driven by restless curiosity, working in a shabby room and taking advantage of the economic value of serendipity [11].

To my knowledge, there is no published history of lithium-ion battery. This may be due to its more recent development and the wider number of professional spheres involved: materials research, electrochemistry, energy and car industries, etc. However, the official narrative might spring from the 2019 Nobel Prize in chemistry awarded to John Goodenough, Stanley Whittingham and Akira Yoshino "for the development of lithium-ion batteries". Their gathering in the same narrative was not obvious. Indeed, each of their Nobel lectures on December 2019 gives an idiosyncratic historical account related to their own techno-scientific interests and professional trajectories. Moreover, a six-page analysis of the history of lithium-ion battery by one STS specialist of the domain mentioned only Goodenough and Whittingham (without Yoshino) with two other groups: Joseph Kummer and co-workers at Ford Motor Company; and Archie Hickling at University College, Leicester [15]. Yet, the symbolic prestige of the Nobel Prize may change the situation by popularizing the following simple and linear account:

> In the early 1970s, Stanley Whittingham used lithium's enormous drive to release its outer electron when he developed the first functional lithium battery. In 1980, John Goodenough doubled the battery's potential, creating the right conditions for a vastly more powerful and useful battery. In 1985, Akira Yoshino succeeded in eliminating pure lithium from the battery, instead basing it wholly on lithium ions, which are safer than pure lithium. This made the battery workable in practice. Lithium-ion batteries have brought the greatest benefit to humankind, as they have enabled the development of laptop computers, mobile phones, electric vehicles and the storage of energy generated by solar and wind power [36].

Another scheme shared by practitioners' narratives on lithium is their linking to systemic mutations, namely "a third revolution" in industry and psychiatry. On the industrial side, the "third industrial revolution" is supposed to follow the steam-engine and coal revolution (1750–1850) and the second one, due to electricity and oil (1870–1930). It refers either to the current era of telecommunication or to the emerging economy based on renewable energy. Lithium-ion battery regulates energy storage for portable devices, electric vehicles and renewable energies as stressed by the 2019 Nobel account. On the psychiatric side, the "third revolution" is supposed to come after a first period when "doctors began to treat the mentally ill as patients rather than criminals" (19th century) and a second one linked to the growing importance of Freud's theory (1880–1950) [19]. It refers to the extensive use of chemical drugs to treat mental illness, including lithium for bipolar disorders after 1950.

These broad periods must be refined in light of historical results. The second part of the 19th century sets the scene for lithium to take on its two social characters. On the medical side, a "system of medical economy" gradually fostered research, development and industrial production towards new "therapeutic agents" based on biological, chemical and physical processes [9, pp. 11–14]. The biomedical approach was strengthened by the "bacteriological revolution" at the end of the century (1880–1900). It coincided, in psychiatry, with the growing importance of a "neuro-physiological approach" that gradually challenged the domination of the alienists (1850–1930) [29, p. 9]. Brain therapy was first dominated by fever treatments and, from 1935 onward, by electroshock treatments [34, p. 8]. Except for a few anecdotal cases, lithium was not used as a psychotropic drug during the 1850–1950 period, but the infrastructural basis of its existence was set. This is also the case for lithium-ion battery. The three features that made electric storage possible and necessary were settled between 1850 and 1930. First, the concept of energy was formalized by thermodynamics (1840–1850). Then, the 1859 lead-acid battery of Gaston Planté became the model for a new kind of battery, labeled "secondary battery", which can be charged and discharged alternatively. Secondary batteries were industrially manufactured by electricians and engineers and sold for the market of electric vehicles [35]. Finally, power grids were built and extended all over Europe and the United States by industrial giant corporations (1880–1930) that are still flourishing nowadays [26].

In both cases, World War II was a transition landmark. For medicine, it boosted the expansion of a medical–industrial complex in Europe and the United States, while health became a central political issue and an economic heavyweight of the mass market [21, pp. 11–15]. Psychiatry was characterized after 1945 by "the closure of mental hospitals and the rise of community care" as well as the increasing use of chemically synthesized psychotropic drugs from the 1950s onward. Two major drugs were tested clinically in 1954: lithium, by Mogens Schou, in the treatment of mania; chlorprozamine, by Jean Delay and Henri Deniker, for its anti-psychotic potential. Psycho-pharmacology increased in the 1950s side by side with the increasing regulatory role of the states and the growing size of the pharmaceutical–industrial complex. The 1970s were marked by the authorization of lithium treatments for bipolar disorders by Food and Drug Administration in the United States. The social regulation gradually shifted "from the cane to the pill" over the 1945–1975 period [10].

State investments and regulations for strategic materials also increased during the Cold War. This framed "a military–industrial–academic complex" for research and development of new advanced materials [31] Solid-state batteries emerged in the late 1960s at the crossroad of academic and industrial stakes: sodium-sulfur battery in 1967 at Ford Motor Company; lithium-ion battery in 1972 at Exxon. These two petro-leum-based companies were exploring the possibility of high energy density battery for electric vehicles. In the 1970s, lithium-based batteries were fostered by the oil crises and environmental critics against petro-leum. Later advancements led to the commercialization of the "lithium–cobalt–oxide battery" by Sony in 1991 and AT&T in 1992. This technology prevailed for energy storage on the mass-market of electronic portable devices such as mobile phones and laptop computers. Strikingly, the 1970s radical political groups had opposite opinions on lithium: while they pushed up lithium-ion batteries through electric vehicles and renewable energy for environmental purposes, the anti-psychiatry movement opposed the extensive use of chemicals, including lithium, in psychiatry.

2. Models and Performances

The development of lithium materials for psychiatry and energy relied on epistemic models and experimental tools. The process was nurtured in the "triple helix of university–industry–government relations" that character-izes "the global knowledge economy" of the second half of the 20th century [16].

Psychiatrists perceive the injection of lithium drugs as a means to regulate the nervous system of their patients and to reduce their pains. Their development and production were driven by the medical–industrial complex worth US$ billions for psychotropic drugs in the 1970s. Chlorpromazine, synthesized by Rhône-Poulenc in the early 1950s and known as Thorazine, was the first anti-psychotic medication: it was used by European and North American psychiatrists to "calm many schizo-phrenic patients without sedating them [...] Increasingly during the 1960s and 1970s, hospital psychiatrists were using a number of phenothiazine derivatives to treat schizophrenia, lithium carbonate to stabilize manic patients and imipramine to relieve psychotic depression" [33, pp.

334–335]. In the 1980s, in Europe and North America, around "one person in about two thousand" regularly took lithium for bipolar disorders [27, p. xiii]. After decades of intensive pharmaceutical R&D to find other "classes of drugs as anticonvulsants and atypical antipsychotics," lithium is still used to regulate "abnormal mood swings" of bipolar patients in the 21st century [5, p. 267].

The simplicity of lithium drugs made the positive ion Li^+ an epistemic model to understand the psychotropic effects of neurotransmitters. Many practitioners assumed that the interactions of lithium-ion with the nervous system would be simpler to characterize than the large organic molecules of other psychotropic medicines: chlorpromazine, reserpine, amphetamine. They tried to substitute lithium for other alkali ions like sodium (Na^+) or potassium (K^+) [27, p. xv]. The assumption of simplicity proved to be empirically wrong.

It is also in its ionic form that lithium works in batteries. Small electropositive ions (H^+, Li^+ or Na^+) were known to move quickly in bi-dimensional structures such as graphite, tungsten bronzes and beta-alumina. This property increased the conduction of electricity in solid-state matter. The idea of using bi-dimensional crystals as cathode materials for advanced batteries was driven by car makers seeking power sources for electric vehicles and by the US Army for light portable electronic devices [2, 43]. It was hoped that solid-state electrolytes would carry more electricity than traditional liquid ones. A NATO conference held in 1972 in Belgirate (Italy) spurred academic and industrial interactions [22]. Cathode and anode materials, which alternatively store ions, posed technological problems. The Stanford Materials Research Laboratory directed by Robert Huggins oriented Michel Armand and Stanley Whittingham toward the use of titanium disulfide for cathode. When he left for Exxon in 1972, Whittingham reused this material to design a workable solid-state battery in the mid-1970s. When Goodenough substituted titanium disulfide for cobalt oxide, the performance of the battery was doubled. Yoshino designed the other side of the battery, the anode, with graphite. During discharge, the lithium-ion left graphite (anode) to intercalate into the welcoming layers of the cobalt oxide, while, during the charge, the electric power forced the lithium-ion to go back and intercalate into the atomic layers of graphite. The reversibility of the discharge–charge cycle was increased by reducing chemical side effects. The 1991 Sony battery, which incorporated a cobalt oxide cathode, a graphite anode and a lithium-based salt as liquid electrolyte, had a specific power of 100 watts per kilogram [38, p. 19]. Its performance has been tripled for three decades but stays at a quarter of its theoretical maximum. Its nominal tension of around 3.6 volts is also the triple of its closer technological competitor, nickel-metal hydride battery. By increasing the quantity of stored energy, lithium-ion batteries increase the autonomy of electronic devices and loosen the meshes of power grids.

The two lithium-based materials share similar traits. Both perform ahead of their commercial competitors but exhibit serious drawbacks. In psychiatry, lithium is used to control symptoms without curing illnesses. In energy, lithium-ion battery is highly inflammable due to the ether-based liquid electrolyte used by Sony. A safer technology based on solid-state electrolytes was developed by Armand with a reduced power [4]. Both materials also required interdisciplinary research. In psychiatry, "pharmacologists, biochemists, physiologists, psychiatrists, psychologists, and many others, joined forces in the search of the century — the search for the key to mental illness" [27, p. xv]. In energy, lithium-ion battery gathered researchers from materials science and engineering, solid-state chemistry, electrochemistry and physics. This might be linked to the fact that both domains required the convergence of instrumental tools (statistics, devices) and theoretical knowledge (models, theories) to stabilize results and provide efficient products [15]. Lithium bridged gaps: between psychiatric phenomena, suggesting "that mania and depression are closely linked conditions and, in all probability, share a common underlying mechanism" [27, p. xvi]; between research fields like "intercalation chemistry" and "solid state ionics". The role of model-material in both domains hides a strong difference in the comprehension of the action of lithium: clear in electrical systems; fuzzy in nervous systems.

3. Global Division of Work

From an economic perspective, lithium supports two global mass-markets. The market of lithium-ion battery is bigger and increases faster with the expansion of portable electronic devices and of electric vehicle markets:

around \$35 billion worldwide in 2019 with a 14% annual increase. By contrast, the US market of lithium psychotropic drugs would be worth around 1 billion dollars nowadays. The quantitative difference lies in the fact that, even if the pharmaceutical industry has a high added value, there are much more laptop computers and cellular phones than bipolar treatments so far in society — in the United States, around 300 million phones and 3 million prescriptions of lithium — and that each electronic device contains more lithium than an annual drug prescription: 1 gram for a cellular phone; 10 grams for a laptop computer; more than 3 kilograms for an electric vehicle; less than 1 micro-gram in one pharmaceutical dose. In its medical life, lithium is so cheap compared to other psychotropic drugs that it is no longer of interest for industry [5, p. 268]. According to Anna Fels, an American psychiatrist, the "grim and undeserved reputation" of lithium as a drug is partly due to the industrial disinterest [18]. She uses the fairy tale of *Cinderella* to balance symbols: "lithium is the Cinderella of psychotropic medications, neglected and ill used" like this poor girl despised by her step-family (i.e. corporations?).

The mass production of lithium-ion batteries is characterized by a harsh economic and political competition to control the international supply chain, from the extraction of raw materials to the refinement of advanced materials and their integration to commercial batteries. Two countries, Australia and Chili, are by far the two first for the extraction of lithium ores [41, p. 178]. Three transnational companies — "the Big Three" — hold more than 50% of the global market of extraction and refinement of lithium materials: Albemarle and FMC Corporation of US origin and SQM of Chilean origin [24, pp. 24–27]. However, the increasing demand of lithium-ion batteries for the energy transition may change the balance of power in a near future. Chile and two of its neighbors, Bolivia, and Argentina, form the "lithium triangle" since they would hold 60% of the globe' known lithium deposits [41, p. 17]. Thanks to the salt flat of Uyuni, a $10,000km^2$ area at 3,600 meters above sea level, Bolivia detains two-thirds of the known world reserves of lithium. At the other side of the chain, China buys around two-thirds of the lithium bulk materials for the production of around half the lithium-ion batteries in the world, the second half being shared by Japan and South Korea. Some business analysts draw a parallel between South America for the extraction of lithium and the Middle-East for that of petroleum, where China would play a similar leading role for lithium as the United States for oil. This is not so obvious for at least three reasons. First, there is no common policy between the three South American countries of the "lithium triangle" and the transnational companies can be stronger than the national states to negotiate the financial conditions of extraction [41, p. 234]. Then, the production costs are relatively low throughout the entire supply chain, the global market of lithium-ion battery being 20 to 30 times larger than that of lithium bulk materials. Last but not least, lithium, although strategical and rare in the short term, is not the most critical element, in the long term, to produce a lithium-ion battery. When considering the ratio reserve/production in the late 2010s, lithium is 2–3 times most critical than rare earths but 10 times less critical than nickel and cobalt [24, p.7].

Thus, the problem is not a geological constraint of ore reserves, but the regulation of the supply chain to get good quality materials in sufficient quantity and at acceptable prices [30, pp. 43–44]. Therefore, Goodenough is still rather "keen on a sodium anode because […] it is available to anyone with access to an ocean, whereas lithium may one day, like oil, require what he calls 'gunboat diplomacy' to secure supplies" [23]. Even if the "gunboat diplomacy" of the early 20th century has not yet returned, the mining competition tends to favor the development of military conflicts, especially in Africa. Working on empiric data covering 92% of African countries and including 14 minerals (but no lithium), geographers established a statistical correlation between mining business and the increase of average violence and spatial extension of military conflicts during the 1997–2010 period, when international prices of raw materials rose due to Chinese and Indian growing economies [6].

4. The Sociopolitical Criticality of Lithium for Energy

Besides economic values, mining generates ecological wastes and social disorders that are all the more dramatic than the states are weaker to negotiate the local conditions of extraction with the transnational companies. The history of Latin America is gashed by many environmental and social problems induced by large-scale mining, as sharply pictured by the iconic essay of Eduardo Galeano: *Open Veins of Latin America: Five Centuries of the Pillage of a Continent* [20]. It tackles the political problem about how lands, works, damages and benefits are distributed between local groups and international powers. Interestingly, Bolivia is "one of the poorest countries" of South America while it has "one of the longest histories of mining" [40]. After the 1985 tin market crash, it turned its "tin-based economy" toward alternative mining businesses (gold, zinc, silver, lead, lithium, potassium) to avoid mine workers joining illegal activities of smuggling and cocaine traffic [14, pp. 84–85]. This national "extractivism" induced strong environmental and social problems in Bolivia, which forced indigenous communities to enforce their territorial rights [39, pp. 102–109].

The left-wing administration of President Evo Morales (2006–2019) used the rise of the international prices of raw materials, especially natural gas, to increase the public investment for the development of mining and industry. Its 2008 creation of a state company, Yacimientos de Litio Bolivianos (YLB), aimed at building the entire supply chain of production in Bolivia since the profit increases through the chain of value, from raw materials to advanced devices. Economic agreements were concluded with transnational corporations based in Asia (China, Japan, South Korea), Europe (France, Germany) and North America [12, p. 93]. If the foreign know-how and investments were needed, the private–public conglomerates were put under public control with 51% shares belonging to the Bolivian state [32]. Such a strategy was intended to increase the state control and collective benefits at the expense of private interests.

The lithium supply chain exhibits the sharp constraints due to the unbalance of power between the ubiquitous transnational corporations, the industrial states that control the production of advanced materials generating high added-value (Asia), and the extractive states where the extraction of bulk materials generate industrial wastes and social conflicts (South America, Africa). A fourth powerful set of actors, Northern America along with Europe, takes advantage of this global division of work by a mass consumption of storage batteries hidden in electronic devices and electrical systems. Indeed, lithium-ion batteries and fuel cells are useful to store energy in electrical systems powered by intermittent renewable energy such as solar and wind powers. They regulate the electrical systems from off-peak to peak hours of consumption and the unpredictable meteorological fluctuations through time and space. This partnership over-sells the image of an ecological transition toward a carbon-free economy where "green" (for renewable) and "blue" (for hydrogen) are valued in contrast to "black" pictures (for oil) [42, p. 71] The tone is even more epic in the 2019 Nobel Prize narrative where the expansion of smart-grids powered by renewable energies is said to work for the "greatest benefit of humankind" [36]. No room is left for the *far-from-Stockholm* environmental and social side effects of the mass extraction of lithium in South America and the mass production of lithium-ion batteries in China. Instead, this politically correct discourse weaves an harmonious symbiosis between fundamental research and industrial profits.

5. Ethical Criticality of Psychotropic Lithium

If lithium-ion batteries contribute to regulate electrical systems, lithium pills did the same for psycho-social systems. Three aspects of lithium regulation will be considered: side effects of medication, the ethics of therapy and the anthropological question of brain regulation. The first issue does not mention social and environmental damages for communities like in the case of lithium mining but biological and psychological effects on individuals. Several effects of lithium have induced "crises" among therapists and patients since the 1950s: thyroid disturbances; renal damages; accentuation of cardiac deficiency; cardiac malformations of fetuses in pregnant patients [27, pp. 123–129]. Psychotropic drugs can also generate nauseous states and sickness in patients. The side-effect issue is related to the antique question of *pharmakon,* in which a drug can have poisonous or therapeutic actions. A good therapy encompasses standards of medication (frequencies, quantities, forms), users' practices (including misuses and over-medication) and patients' idiosyncrasies (individuality, history). As the negative side effects are usually recorded to produce improved drugs [37], it could be expected that the epistemic gap between the statistical results and the patients' individual experience would be overcome [13, p. 648]. However, the lithium effects on sample populations during 50 years of clinical trials [25, pp. 534–536] hardly converged with the trial-and-error adjustment for a "long-term or even life-time medication" to get a suitable therapeutic performance.

This epistemic discrepancy questions the ethical values of pharmaceutical treatments to regulate and control the social behaviors of individuals. Lithium may contribute to "the intensified standardization of human beings": a century after the industrial standardization of products such as batteries, the individual is threatened to become an *homo artefact* through mass-market series [8, pp. 142–144].

Besides personal and ethical impacts, the chemical normalization of psychiatry led to a third set of issues at the anthropological level. What becomes of the individual when he is deeply regulated by psychotrops? "In such a medicalized culture, diagnoses make sure that individuality is left in the dark — and along disappears the individual's struggle for freedom and autonomy" [8, p. 142]. The individual is also the one who acts, thinks or feels, remembering, refusing or hoping for his evanescent self. To what extent would it be possible to break out of the regulated box under lithium?

Lithium, don't want to lock me up inside.
Lithium, don't want to forget how it feels without...
Lithium, I want to stay in love with my sorrow.
Oh, but God, I want to let it go [17].

6. Conclusion

A ubiquitous element, lithium reached an academic status in 19th century chemistry labs and textbooks before embarking on two parallel careers in 20th century consumer societies: energy and psychiatry. In both domains, it climbed the social ladder to advanced materials thanks to the joint efforts of industrial complexes, research laboratories and state policies. It became strategic for psychiatric treatments in the 1970s and for energy storage in the 1990s. In spite of defects and risks, its small, interactive, boundary-crossing ionic form makes it a world leader to treat bipolar patients and to energize electronic devices and electric vehicles in the early 2020s. This was due to its great ability to regulate nervous and electrical systems according to managerial and economic standards. On the social side, lithium was criticized for ethical reasons. It was enrolled under the green banner of the energy transition toward a carbon-free utopia, whereas its large-scale mining provoked environmental damages and serious social disorders. Its psychotropic use provided relief to some human sufferings by making individuals less energetic and its therapeutic extension hardly hid the growing psycho-social pressures of capitalist competitions. Finally, lithium-based materials collaborated toward the normalization of individuals on the psychiatry side and the flexibilization of electric systems on the energy side.

References

1. Abir-Am, P. and Elliot, C. E. (1999). *Commemorative Practices in Science: Historical Perspectives on the Politics of Collective Memory*. Chicago: University of Chicago Press.
2. Armand, M. (September 18, 2001). Interview by Bensaude-Vincent B, Hessenbruch A. Sciences: Histoire orale. https://sho.spip.espci.fr/spip.php?article8.
3. Bach Ricardo, O. and Gallicchio Vincent, S. (Eds.) (1990). *Lithium and Cell Physiology*. New York, Berlin, Heidelberg, London, Paris, Tokyo, Hong Kong: Springer-Verlag.
4. Barboux, P. (December 12, 2000). Interview by Bensaude-Vincent B. Sciences: Histoire orale. https://sho.spip.espci.fr/spip.php?article46.
5. Bech, P. (2006). The full story of lithium. *Psychotherapy and Psychosomatics*. 75(5), 265–269.
6. Berman, N., Couttenier, M., Rohner, D. and Thoenig, M. (2017). This mine is mine! How minerals fuel conflicts in Africa. *The American Economic Review*. 107(6), 1564–1610.
7. Berzelius, J. J. (1817). Ein Neues Mineralisches Alkalin und ein neues Metall. *Journal für Chemie und Physik*. 21, 44–48.
8. Birkholm, K. (2016). The ethical judgment: Chemical psychotropics. *HYLE — International Journal for Philosophy of Chemistry*. 22, 127–148.
9. Bonah, C. and Rasmussen, A. (2005). *Histoire et médicament aux XIXe et XXe siècles*. Paris: Biotem & Éditions Glyphe.
10. Bray, A. (2009). Chemical-Control^{TM}®: From the Cane to the Pill. In Poster, M. and Savat, D. (Eds.) *Deleuze and New Technology*. Edinburgh: Edinburgh University Press. pp. 82–103.
11. Brown, W. (2019). *Lithium: A Doctor, a Drug, and a Breakthrough*. New York: Liveright Publishing.
12. Brugger, F. and Zamora, K. L. (2014). Why commodity booms have not (Yet?) boosted human capital: Bolivia's struggle to create a skilled workforce. In Carbonnier, G., Carton, M., and King, K. (Eds.) *Education, Learning, Training*. Critical Issues for Development. Brill. Chapter 5, pp. 81–101. Leiden: Brill Publishers.

13. Busfield, J. (2003). Mental illness. In Cooter, R. and Pickstone, J. (Eds.) *Companion to Medicine in the 20th Century*. London & New York: Routledge. Chapter 41, pp. 633–651.

14. Crabtree, J., Duffy, G. and Pearce, J. (1987). Bolivia: The end of the tin era? The Great Tin Crash. Bolivia and the World Tin Market. Latin American Bureau. pp. 83–91.

15. Eisler, M. N. (2020). Power sources. In Martin, J. D. and Mody, C. C. M. (Eds.) *Between Making and Knowing. Tools in the History of Materials Research*. Singapore: World Scientific. Chapter 3.8, pp. 349–355.

16. Etzkowitz, H. and Leydesdorff, L. (Eds.) (1997). *Universities and the Global Knowledge Economy: A Triple Helix of University-Industry-Government Relations*. London: Pinter.

17. Evanescence (Amy Lee). (2006). Lithium. In *The Open Door*. Wind-up. Refrain.

18. Fels, A. (September 13, 2014). Should we all take a bit of lithium? *The New York Times*.

19. Fieve, R. R. (1975). *Moodswing: The Third Revolution in Psychiatry*. New York: William Morrow and Company.

20. Galeano, E. (1971). *Las venas abiertas de América Latina*. Colombia: Siglo XXI Editores.

21. Gaudillière, J-P. (2002). *Inventer la biomédecine: la France, l'Amérique et la production des savoirs du vivant (1945–1965)*. Paris: La Découverte.

22. Gool, W. van (Ed.) (1973). *Fast Ion Transport in Solids: Solid State Batteries and Devices*. Amsterdam: Elsevier.

23. Goodenough, J. (November 29, 2018). Brain scan. The Economist, Technology Quaterly.

24. Hache E, Simoën M, Seck GS. Electrification du parc automobile mondial et criticité du lithium à l'horizon 2050. Paris: Ademe Report; 2018.

25. Hayward, R. (2011). Medicine and the mind. In Jackson, M. (Ed.) *The Oxford Handbook of the History of Medicine*. Oxford: Oxford University Press. Chapter 29, pp. 524–542.

26. Hughes, T. P. (1993). *Networks of Power: Electrification in Western Society, 1880–1930*. Baltimore: Johns Hopkins University Press.

27. Johnson, F. N. (1984). *The History of Lithium Therapy*. London: MacMillan.

28. Lukasz, L. (2014). Six phases of cosmic chemistry. *HYLE — International Journal for Philosophy of Chemistry*. 20, 165–192.

29. Le Jeune, K. (2014). Une histoire de l'épilepsie aux XIXe et XXe siècles. Définition et développement d'une pathologie entre neurologie et psychiatrie. Ph.D., University of Nantes.

30. Lepesant, G. (2018). *La transition énergétique face au défi des métaux critiques*. Paris: Études de l'IFRI.

31. Leslie, S. W. (1993). *The Cold War and American Science: Military-Industrial-Academic Complex at MIT and Stanford*. New York: Columbia University Press.

32. Mariette, M. (January 2020). En Bolivie, la filière lithium à l'encan. Le Monde diplomatique. 23.

33. Micale, M. S. (2003). The psychiatric body. In Cooter, R. and Pickstone, J. (Eds.) *Companion to Medicine in the 20th Century*. London & New York: Routledge. Chapter 22, pp. 323–346.

34. Missa, J-.N. (2006). *Naissance de la psychiatrie biologique : histoire des traitements des maladies mentales au XXe siècle*. Paris: Presses universitaires de France.

35. Mom, G. (2004). *The Electric Vehicle: Technology and Expectations in the Automobile Age*. Baltimore: The Johns Hopkins University Press.

36. Nobel Prize. (2019). https://www.nobelprize.org/prizes/chemistry/2019/popular-information/.

37. Parisi, L. and Terranova, T. (2000). Heat-death: Emergence and control in genetic engineering and artificial life. CTheory. A084. http://www.ctheory.net/text_file.asp?pick=127.

38. Pistoia, G. and Liaw, B. (Eds.) (2014). *Behaviour of Lithium-Ion Batteries in Electric Vehicles. Battery Health, Performance, Safety, and Cost*. Cham: Springer.

39. Postero, N. (2017). Living well? The battle for national development. In *The Indigenous State. Race, Politics, and Performance in Plurinational Bolivia*. Oakland: University of California Press.

40. Santiago, M. (2013). Extracting histories: Mining, workers, and environment, Rachel Carson center perspectives. Vol. 7. New Environmental Histories of Latin America and the Caribbean. pp. 81–88.

41. Serandour, A. Le "triangle du lithium" à l'heure globale : marges et intégrations territoriales (Argentine, Bolivie, Chili). PhD in geography, University of Paris, 2020.

42. Teissier, P. (2017). From the birth of fuel cells to the Utopia of hydrogen world. In Bensaude-Vincent, B., Loeve, S., Nordmann, A. and Schwarz, A. (Eds.) *Research Objects in Their Technological Setting*. London & New York: Routledge. Chapter 5, pp. 70–86.

43. Whittingham, M. S. (October 30, 2000). Interview by Bensaude-Vincent B, Hessenbruch A. Sciences: Histoire orale. https://sho.spip.espci.fr/spip.php?article132.

Chapter 15

Rare Earths

Soraya Boudia

University of Paris, France

1. Introduction

The history of rare earths is one of old, 19th century elements that have acquired major economic and political importance in the past fifteen years [1, 11, 19, 25]. According to the definition of IUPAC (International Union of Pure and Applied Chemistry), rare earths are a family of 17 metallic elements on the periodic table: 15 lanthanides (lanthanum, cerium, praseodymium, neodymium, promethium, samarium, europium, gadolinium, terbium, dysprosium, holmium, erbium, thulium, ytterbium and lutetium) plus candium and yttrium. Their unique magnetic, luminescent and catalytic properties have fascinated chemists and led to a variety of industrial applications [2].

Rare earths were long unknown to the general public, their names only meaningful to scientists, engineers and metals traders. They were forced onto the public scene in 2010–2011 during what was called the rare earths crisis [4]. At the heart of this crisis was China, which has a near monopoly over the extraction and primary mineral processing industry, producing 80–90% of the rare earth oxides destined for a variety of industries around the world. The explosion of their cost on international markets and the ensuing concerned public discourses about a possible exhaustion of supply reveal their strategic importance. Their importance is not due to the volume that is traded, which is limited in comparison to the production of the better-known metals like iron, copper and aluminum. This gap is tellingly summed up in one figure: the total production of rare earths amounts to barely 0.01% of steel production [27]. Rare earths are important because of their use in several strategic and military sectors: the electronic industry, telecommunications and renewable energy, that is in nearly all contemporary technological products. Some engineers and economists call them the "elements of modernity," the "vitamins" or "hormones" of the economy.

This chapter shows how rare earths reveal industrial modernity in all its ambiguity. It first examines how practical and theoretical knowledge of these materials with remarkable properties has made it possible to develop many applications that have profoundly shaped contemporary society, be it lighting, color television, smartphones or computers. It then follows the transformation of their status as critical materials following the 2010–2011 crisis. Material criticality is an approach based on the evaluation of risks related to the production, use and management of end use of a primary material [14]. Many countries have declared rare earths critical because of their role in several strategic industries (energy, defense, telecommunications), even if the reserves and production are concentrated in China and the supply chain is vulnerable to breakdown. The murkiness of industrial modernity manifests itself in three ways. First, it is materialized in the organization of the worldwide value chain that generates risky interdependencies, even for materials considered to be strategic [12,16]. It secondly appears in the heavy environmental and human costs of the extraction of rare earths and the chemical pollution and waste of their production. Due to the international division of production, this cost weighs most heavily on the Chinese population, with the toxic legacy of past rare earth production in various parts of the planet and the ongoing mining boom contributing greatly to accentuating the global environmental crisis. Lastly, the darkest side of our intercourse with rare earths lies in a prickly paradox: the ecological and energy transition that is promoted by so many actors today is based on an assortment of low-carbon technologies (electric cars, wind turbines, digital technologies) that require ever-increasing quantities of rare earths, contributing to the intensification of their extraction and production, generating major pollution that is as much a burden today as for future generations.

2. A Long History of Industrial Applications

The chemist Johann Gadolin introduced the term "rare earths" in 1794, to describe oxide of ytterbite (extracted through a series of chemical treatments), discovered in 1787 by Carl Axel Arrhenius, a Swedish army officer and amateur geologist and chemist [22]. The term "earths" was a reference to an old name for the insoluble residues left after combustion or a long string of chemical treatments [30]. The term "rare" initially referred to the difficulties in separating these earths from their ores; rare earths are not as rare as their name suggests because they are relatively abundant on the surface of the Earth and can be extracted from deposits of many minerals, especially monazite and bastnäsite.

The history of the discovery of various members of the rare earth family is concentrated in a period of 150 years, marked by a number of sometimes lively debates, particularly over difficulties in isolating them. In 1803, cerium was identified by Martin Heinrich Klaproth in Germany and by Jöns Jacob Berzelius and Wilhelm Hisinger in Sweden, and in 1947 the discovery of the last element of the family, promethium, was announced, isolated by Jacob Marinsky and Lawrence Elgin Glendenin while working on the Manhattan Project [23]. Knowledge on rare earths accumulated through constant back-and-forth between research labs and industrial workshops in a combination of interests in isolating new elements to fill in Mendeleev's table, determining and understanding the properties of these elements, and developing their industrial utility through various technical innovations.

The scientific and industrial history of rare earths took off in the last third of the 19th century. It was marked by contributions from many figures in chemistry: Carl Gustaf Mosander of Sweden, Auer von Welsbach of Vienna, Paul-Émile Lecoq de Boisbaudran and Georges Urbain of France, who contributed to the discovery of several elements (cerium, lanthanum, neodymium, praseodymium, lutetium, samarium…) [8]. The refinement of a series of applications gradually made these elements indispensable to several sectors. The first was the use of rare earths for urban illumination. In 1885, the rare earth extraction industry began to grow with the production of a model of wildly successful gas mantles, the Auer mantle. Working with Robert Wilhelm Bunsen and Gustav Kirchhoff, Auer Von Welsbach introduced various oxides into a Bunsen burner to bring them to incandescence. He anticipated the importance of his discovery and filed several patents. The first was for the use of zirconium oxide, but then Auer realized that thorium gave better results for incandescence. The blend was 99% thorium dioxide and 1% cerium dioxide. The resulting light was much brighter than that of gaslight in use at the time, while consuming less gas. After World War I, lighting using gas wicks was replaced with electric lighting. Production continued, however, to serve other parts of the world, Southeast Asia in particular.

The use of rare earths boomed again with the development of alloys, the best known being mischmetal (from the word for "mixed metal" in German), whose basic composition is 45–50% cerium, 25% lanthanum, 15–20% neodymium and 5% praseodymium. Its history is rooted in the desire to make use of industrial waste from the production of gas mantles. Rare earths were extracted from monazite ore, which was processed for its thorium (containing 5–9% thorium oxide) while 60% of the residue was considered waste. Auer Von Welsbach discovered that by filing certain alloys of rare earths, he could create bright sparks that would light combustible materials (petroleum or tinder at the time). In 1903, he patented lighting flint, an alloy composed of 70% mischmetal and 30% iron, and for a time had a monopoly over the global pyrophoric alloy market.

The growth of rare earth industry was dependent on the mining of mineral deposits and their organized international circulation. In the beginning the European industry was reliant on ores from India and Brazil. The 1949 discovery of a bastnäsite deposit at Pass Mountain, in the Californian Mojave desert, shifted the center of gravity of the rare earths industry from Europe to the United States. Mining began in 1952 and grew rapidly thanks to the industrialization of cathode-ray tubes for color televisions. Four rare earth elements — europium, yttrium, thulium and terbium — are behind a technical innovation that revolutionized both the home and the public sphere. When these elements are struck by an electron beam going through the tube, they produce the primary colors of light (red, green and blue) by phosphorescence, creating the pixels of the images on those old fashioned CRT televisions.

After their use in making lighter flints, creating simple glass and ceramic pigments, and coloring television, several rare earths were sought out for their magnetic and semi-conductive properties. Today they are indispensable components in digital technologies and the green energy industry, notably due to the use of neodymium, dysprosium and samarium in magnets. Compared to classic magnets made of ferrite, rare earths make it possible to produce magnets a tenth of the size for the same power [35]. Such properties allow miniaturization and increased calculation power, fostering the development of many objects with computing capacities, including smartphones.

The expansion of the use of rare earths is directly reflected in the increase in the volume of their production. Prior to the 1960s, this production of rare earth oxides was estimated at 10,000 metric tons per year. Global demand grew significantly with the communication technology and micro-electronics boom, which helped push production up to 53,000 tons in 1990, 90,000 tons in 2000, to reach an estimated volume of 142,000 to 159,000 tons by 2010 [6, 29, 32].

Rare earths are now indispensable to many industries, telecommunications, petroleum and renewable energies chief among them. Over the past 15 years, they proved to be so important that they have been put on the list of so-called critical or strategic materials, a situation that can lead to notable geopolitical tensions, primarily with China as it monopolizes production.

3. A Sudden Crisis

This critical status changed following the so-called "rare earths crisis" in 2010–2011, which revealed the scope of the Chinese production monopoly. China built up its monopoly on mining the Bayan Obo deposit in inner Mongolia. The Pass Mountain mine had been the largest site in the world from the 1960s through the 1990s, but Molycorp decided to phase out its activity in the United States and transfer it to China, as part of the reorganization of its entire production chain [13]. Starting in the 1980s, Edward Nixon (younger brother of US president Richard Nixon) approached the upper management of Unocal, of which Molycorp was a subsidiary, offering to contract part of its operations in China for economic reasons as well as to address the "environmental problems" that I will expand on shortly. Nixon's firm facilitated the transfer of several operations by establishing joint ventures with the Chinese [34]. In 1998, Molycorp began to phase out its refining activities in the US and export thousands of tons of the ore it was still mining to China for processing, only to ship it back in the form of oxides for various industrial sectors.

Possessing half of the known reserves of rare earths, China henceforth ensured 85–90% of world production. Given this monopolistic position, it tried to renegotiate its position in the world of rare earths, which led to a commercial, financial and geopolitical crisis. Two episodes were decisive in setting it off. The first began in 2005, when China announced its intention to establish biannual export quotas for rare earths. In 2009, China exported 50,145 tons and announced a drop in the quota for the following year, to 22,282 tons the first semester and 7976 the second, amounting to an overall drop of 40% for 2010. It additionally established new export taxes of 15% to 20%, according to the type of metal.

Act two occurred in September 2010, further exacerbating concerns raised by the announcement of these quotas. It took the form of a diplomatic crisis between China and Japan in a long-running territorial dispute over the Senkaku (or Diaoyu) Islands. Tensions spiked that September following the arrest of the captain of a Chinese fishing trawler that had collided with a Japanese coast guard patrol near the islands. Japan then announced that in retaliation China had decided to stop all exports of rare earths to Japan, making a powerful geopolitical weapon of these elements that were indispensable to the Japanese high-tech economy. This led to an increase in purchasing on the global market and soaring prices. The rise in market prices between September 2010 and August 2011 was spectacular. Most prices doubled, and a few rose even more dazzlingly: the price of dysprosium oxide rose by 1100% and neodymium by 1000% [34].

Skyrocketing prices, fears of exhausting the supply and concern about the growing dependence on China, all these symptoms of "crisis" prompted an international mobilization. While a group of actors from Japan, the US and Europe developed public and private initiatives, alarmist discourses proliferated about shedding light on the geopolitical and commercial risks tied to securing strategic resources for various civil and military sectors. A profusion of news articles came out reporting "tensions as never seen before," a "metals war," and the possibility of exhausting the supply of metals essential to strategic industrial and military sectors [5]. Analysts who spoke to the media attributed responsibility for the situation to China's increasingly manifest desire for economic and political hegemony and certain Euro-American policies that had failed to act in anticipation of China's strategy, which had been announced years earlier. They called for the development of ambitious policies to bring solutions to what they considered to be a particularly alarming situation.

A number of public and private initiatives developed because of the major risk to the security of the strategic industrial sectors of defense, telecommunications and energy [26, 28]. In late 2010, the Japanese government, concerned about the impact of the new situation on its electronics industry, announced the establishment of a special fund of 100 billion yen ($1.2 billion) in support of its industrial sectors using rare earths at a large scale, headed by Mitsubishi. Part of this sum could be used to seek new ore deposits outside of China, to develop new extraction and recycling procedures, and to stockpile rare earths to buffer the risk of supply chain interruption [15]. In Europe, the European Commission adopted a similar approach, seeking to establish a consistent strategy for so-called critical materials considered indispensable to the most important economic sectors [7, 10].

The United States was a key actor in this time of crisis, since American industry consumed a significant share of rare earths from China. The Center for Strategic and International Studies, a think tank well known for its work on international politics and national security, held a conference in December 2010 assembling official actors including the US Department of Energy, the United States Geological Survey, industry representatives, the think tank's own experts in energy and national security, and a few consultants — many of them former upper management from the metals sector reputed for their expertise. At the same time, the Department of Energy published an initial report [31] that emphasized in a measured way that the American economy and national security could be severely impacted in the future by a lack of a steady supply of rare earths, and that it would take 15 years to free the United States from its dependency on the supply coming from China. It thus recommended establishing a national strategy for so-called critical materials in three parts: intensifying research, increasing diplomatic contacts to find new sources, and developing a policy for recycling and replacing rare earths. In 2010, Congress granted it an additional budget of $15 million to bolster the activities of its research departments and $35 million for work on the next generation of batteries without rare earths. The Department of Energy developed a research and development plan for critical materials, funded the first research into what were deemed priority themes, and launched a series of meetings of international experts on the subject. Jointly with the National Science and Technology Council (NSTC) it helped to create the Subcommittee on Critical and Strategic Mineral Supply Chains in order to facilitate the coordination of the activities of various federal agencies concerning strategic material supplies.

In this climate of geopolitical crisis, experts and policy-makers developed strategies to face a situation of emergency while also anticipating the future. It was a matter of finding the means to put pressure on China so that it would reconsider its decision, on the one hand, and reconsider national production and supply policies for strategic materials, develop research and development, promote new technologies for recycling and finding replacements, and spur prospecting for ore in countries other than China, on the other. One of the important initiatives to find a resolution is the complaint that the United States, the European Union and Japan lodged with the World Trade Organization (WTO) in 2012, accusing China's action of violating WTO trade regulations. The plaintiffs won their case in March 2014. China lifted its quotas in January 2015, and export taxes on 2 May 2015, although the market prices of rare earths had been declining since 2012 [18]. This drop in price and the obligation for China to bring exports back to earlier levels put an end to the crisis.

4. An Unsustainable Industry for the Ecological Transition

This five-year crisis had the effect of drawing the media attention on an economic sector that was little known to the public at large. Scores of journalists helped to document the production conditions of these strategic elements. Taken altogether, the full range of accounts that they reported paints an image that some would unhesitatingly describe as apocalyptic.

This is especially true for reporting on the Inner Mongolian city of Baotou, the world's rare earths capital, where most factories processing minerals from the Bayan Obo deposit 120 km north are located.

All accounts converge on descriptions of the ravages of rare earth extraction and processing. BBC journalist Tim Maughan has described the mining site as "hell on earth" [25] and revealed massive pollution from wastewater discharge, gas emissions and solid toxic waste, not to mention radioactive waste from ore processing.

This major pollution is not only due to China's attitude of environmental dumping to lower production costs and thus dominate the world market. As many studies have documented, it also results from the very conditions of rare earths' production, due to their properties. The extraction and separation of these metals that are present in ores in small quantities require hydrometallurgical procedures on an industrial scale and repeated use of acid baths [24]. This processing produces toxic sludge that seeps into the environment or is pumped into fields of concrete "ponds" in sites that can be several square kilometers. Acidic water seeps from these ponds and other waste storage facilities, altering local soil pH and carrying heavy metals and other toxic substances, including lead, arsenic and radioactive thorium and uranium. This has led to massive soil and groundwater contamination well beyond the mining site, and vast zones are now subject to generalized poisoning.

While many critics alerted about this pollution, the environmental dimension is nearly absent from analyses during discussions on the rare earths crisis. Not once was the crisis described as environmental or ecological, even though China proclaimed it at every opportunity. A sizable body of Chinese research has documented the scale of chemical and radioactive contamination in regions of rare earth extraction and processing and their health effects on workers and nearby nomadic and sedentary populations [20]. One of the major concerns is the contamination of the water supply by toxic chemicals — especially heavy metals — whose health effects have gotten so serious that Chinese authorities have banned the consumption of water in affected regions [2, 17]. Authorities also had to respond to the mobilizations of villagers subjected to this large-scale pollution who turned to the central government to demand environmental oversight, compensation for damages and relocation for all concerned households in 2004. These problems led many Chinese experts and leaders to highlight the environmental cost that their country pays to supply the world with metals of strategic importance for a growing number of industries. China has indicated that it can no longer ignore the environmental cost of this production and has requested that this burden be shared. It called on other countries to (re)open mines to assume their share of pollution flowing from digital technologies and renewable energies. This demand was an opportunity to remind the world that it only held 50% of ore deposits but ensured 80% of global supply.

Described in terms of a crisis, this turbulent episode in the history of rare earths sheds a light on a chronic problem: the production conditions of these metals and the ensuing pollution in the countries of production reveals the environmental costs of the growth of contemporary technologies. On might conclude that this crisis had the benefit of exposing and publicizing the environmental damages of these extractive activities. But when the issue is considered over the long term, it is evident that the environmental cost of rare earths production is not entirely new. As a matter of fact, it already changed the geography of rare earth production over the years leading up to this crisis, but this aspect was nearly invisible in the 2010–2011 crisis. Indeed, the cost of environmental pollution from rare earth production was a major factor in decisions to close several mining and industrial sites in many Western countries and offshore a large share of extraction and refining to China.

In the United States, Molycorp's decision to halt rare earth production was part of a broader trend to offshore many mining and manufacturing activities to China due to the lower cost of production and less restrictive regulations there. The decision was the consequence of the many problems they had encountered due to major pollution caused by extraction and refining. The company had to deal with a string of incidents, most seriously repeated leaks in the partly underground pipeline bringing wastewater from the mine to a settling pond 22.5 km (14 miles) away. The legal consequences of these leaks of radioactive water full of toxic products were costly for the business and spurred the end of rare earth production on the site. The United States Environmental Protection Agency (EPA) was relatively slow to regulate mining activities, only starting to do so in the late 1970s. Under pressure from environmentalists, the EPA conducted a retrospective evaluation of pollution engendered by Mountain

Pass mining activities that revealed several occasions where such wastewaters were dumped. By 1990, over 45,000 gallons of polluted water had been dumped, some of it at the boundary of the Mohave National Reserve and near schools and homes [9]. The company had to face legal trouble and virulent criticism from environmentalists while also dealing with the drop in the price of rare earths due to Chinese production. By offshoring its extracting and mining activities, Molycorp also off-shored the pollution.

The crisis can be read as a moment of renegotiation when China sought to change its position in the international chain, from site of basic extraction and production to producer of high value-added objects. A number of initiatives and attempts to break China's monopoly followed this crisis, although so far to little effect. The production of rare earth oxides coming out of China rose to about 120,000 tons in 2018, out of a worldwide production of 170,000, amounting to over 70% of the global total; Australia is the second producer, extracting only 20,000 tons (12%), the United States 15,000 tons (9%), Myanmar 5,000 tons (3%) and Russia 2600 tons (1.5%) [33]. New production practices are promoted, but the burden of past and future pollution remains, with countless consequences, especially for future generations.

References

1. Abraham, I. (2011). Rare earths: The cold war in the annals of travancore. In Hecht, G. (Ed.) *Entangled Geographies: Empire and Technopolitics in the Global Cold War*. Masschusetts: MIT Press. pp. 101–124.
2. Ali, S. H. (2014). Social and environmental impact of the rare earth industries. *Resources*. 3, 123–134.
3. Balaram, V. (2019). Rare earth elements: A review of applications, occurrence, exploration, analysis, recycling, and environmental impact. *Geoscience Frontiers*. 10(4), 1285–1303.
4. Boudia, S. (2019). Quand une crise en cache une autre: la 'crise des terres rares' entre géopolitique, finance et dégâts environnementaux. *Critique Internationale*. 85(4), 85–103.
5. Bradsher, K. (2010 September 22). Amid tension, China blocks vital exports to Japan. Last consultation 2021 February 8. https://www.nytimes.com/2010/09/23/business/global/23rare.html.

6. Bureau de recherches géologiques et minières (BRGM). (2015). *Panorama 2014 du marché des Terres Rares*. Orléans: BRGM.

7. Catinat, M. and Anciaux, P. (2011). L'Union européenne et les minerais stratégiques. *Géoéconomie*. 59(4), 44–51.

8. Claro-Gomes, J. M. (2003). *Georges Urbain (1872–1938): chimie et philosophie*. PhD Dissertation. Paris: University of Nanterre.

9. Environmental Protection Agency. (2013). *Rare Earth Elements: A Review of Production, Processing, Recycling, and Associated Environmental Issues*. Cincinnati, OH: Office of Research and Development.

10. European Rare Earths Competency Network. (2015). *Strengthening of the European Rare Earths Supply Chain. Challenges and Policy Options*. Bruxelles: Commission européenne.

11. Evans, C. H. (1996). *Episodes from the History of the Rare Earth Elements*. Dordrecht, Boston and London: Kluwer Academic Publishers.

12. Gereffi, G. and Korzeniewicz, M. (Eds.) (1994). *Commodity Chains and Global Capitalism*. Westport CT: Greenwood Press

13. Goldman, J. A. (2014). The U.S rare earths industry: Its growth and decline. *Journal of Policy History*. 26, 139–166.

14. Graedel, T. E. and Nuss, P. (2014). Employing considerations of criticality in product design. *Journal of the Minerals Metals & Materials Society*. 66(11), 2360–2366.

15. Hatakeyama, K. (2015). Rare earths and Japan: Traditional vulnerability reconsidered. *In Kiggins*. 43–61.

16. Hopkins, T. K. and Wallerstein, I. (1986). Commodity chains in the world economy prior to 1800. *Review*. X(1), 157–170.

17. Huang, X., Zhang, G., Pan, A., Chen, F. and Zheng, Ch. (2016). Protecting the environment and public health from rare earth mining. *Earth's Future*. 4, 532–535.

18. Humphries, M. (2013). *Rare Earth Elements: The Global Supply Chain*. Washington, D. C.: Congressional Research Service.

19. Kiggins, R. D. (Ed.) (2015). *The Political Economy of Rare Earth Elements: Rising Powers and Technological Change*. Londres: Palgrave McMillan.

20. Klinger, J. M. (2015). The environment-security nexus in contemporary rare earth politics. *In Kiggins* [note 19]. 133–155.

21. Klinger, J. M. (2015). A historical geography of rare earth elements: From discovery to the atomic age. *The Extractive Industries and Society*. 2(3), 572–580.

22. Kragh, H. (1996). Elements no 70, 71 and 72: Discovreries and controversies. *In Evans* [note 11]. 67–89.

23. Marinsky, J. A. (1996). The Search for Elements 61. *In Evans* [note 11]. 91–107.
24. Martin, A. and Iles, A. (2020). The ethics of rare earth elements. Over time and space. *HYLE — International Journal for Philosophy of Chemistry*. 26, 5–30.
25. Maughan, T. (2015 April 2). The dystopian lake filled by the world's tech lust. (Accessed 2018 September 9). http://www.bbc.com/future/story/20150402-the-worst-place-on-earth.
26. Parthemore, Ch. (2011). *Elements of Security: Mitigating the Risks of U.S. Dependences on Critical Minerals*. Washington, DC: Center for a New American Security.
27. Pitron, G. (2018). *La guerre des métaux rares. La face cachée de la transition énergétique et numérique*. Paris: Les liens qui libèrent.
28. Resnick Institute for Sustainable Energy Science. (2011). *Critical Materials for Sustainable Energy Applications*. Pasadena, CA: Resnick Institute for Sustainable Energy Science.
29. Roskill Information Services. (2015). *Rare Earths: Market Outlook to 2020*. London: Roskill.
30. Tansjö, L. (1996). Carl Gustaf Mosander C. G. and his research on rare earths. *In Evans* [note 11]. 37–54.
31. U.S. Department of Energy, Critical Materials Strategy. (2010). Washington: U.S. Department of Energy.
32. U.S. Geological Survey (USGS). (2015). *Mineral Commodity Summaries*. Reston, VA: USGS.
33. U.S. Geological Survey. (2019 February). Mineral commodity summaries. (Accessed 2021 February 28). https://prd-wret.s3.us-west-2.amazonaws.com/assets/palladium/production/atoms/files/mcs-2019-raree.pdf.
34. Wübbeke, J. (2015). China's rare earth industry and end-use: Supply security and innovation. *In Kiggins* [note 19]. 20–42.
35. Zept, V. (2013). *Rare earth elements: A New Approach to the Nexus of Supply, Demand, and Use exemplified along the Use of Neodymium Permanent Magnets*. Berlin, Heidelberg: Springer-Verlag.

Index

www.ingramcontent.com/pod-product-compliance
Lightning Source LLC
Chambersburg PA
CBHW050749150726
48196CB00004B/391